Origins in Acoustics

Origins in Acoustics

The Science of Sound
from Antiquity to the Age of Newton

Frederick Vinton Hunt
with a Foreword by Robert Edmund Apfel

New Haven and London
Yale University Press
1978

534.09
H939
C. 2

Designed by Thos. Whitridge and set in IBM Century
type. Printed in the United States of America by
LithoCrafters, Inc., Chelsea, Michigan.

Published in Great Britain, Europe, Africa, and Asia
(except Japan) by Yale University Press, Ltd., London.
Distributed in Australia and New Zealand by Book &
Film Services, Artarmon, N.S.W., Australia; and in
Japan by Harper & Row, Publishers, Tokyo Office.

Library of Congress Cataloging in Publication Data

Hunt, Frederick V.
    Origins in acoustics.

    Bibliography: p.
    Includes index.
    1. Sound—History. I. Title.
QC221.7.H86   534'.09   78-5032
ISBN 0-300-02220-4

Sound, that Noble **Accident** of the **Air** . . .

Saggi . . . Accademia del Cimento
Waller translation

CONTENTS

FOREWORD

Frederick Vinton Hunt died in April 1972, shortly
after his sixty-seventh birthday. He died suddenly at a meeting of the Acoustical Society of America where he had
vigorously led a session on physical acoustics the afternoon
he was taken ill.

When he chose the title *Origins in Acoustics* for the manuscript left unfinished at his death, he could not have realized
how appropriate it would be. Friends and family who went
through his papers found over 1,000 pages of typescript, in
varying stages of completion, covering the history of acoustics
from antiquity to the mid-twentieth century. A close review
of the material led several of us to agree that the early sections, completed in 1958–1959 and bringing the history up to
the Age of Newton, represented a finished draft, written and
rewritten, crafted and polished in a style characteristic of
Hunt. Of the four chapters in the full manuscript—"Origins
in Observation," "Origins in Experiment," "Origins in
Theory," and "Origins in Control and Exploitation"—the
present book comprises the first two plus an appendix made
up of the finished portion of the third. All of this material—
text and endnotes—appears here as Hunt had prepared it,
without substantive editing.

The rest of chapter 3 and all of chapter 4 of the original
manuscript were far from finished (the last draft is dated
1951). Furthermore, Hunt was bothered that his history of
the twentieth century had a different flavor from the rest of

the manuscript; he feared his objectivity might be clouded by his more intimate knowledge of this period. There were also some major gaps in acoustical history. For instance, Lord Rayleigh's unsurpassed contributions to the science of acoustics in the nineteenth and early twentieth centuries were largely omitted because Hunt was well aware of R. Bruce Lindsay's excellent introduction to the Dover edition of Rayleigh's *Theory of Sound.* Hunt himself had written a ninety-one-page introduction to his book *Electroacoustics* which traced the enormous progress in this area between the eighteenth and twentieth centuries, and he therefore did not carry this information over to his historical manuscript. For the seventeenth century, much of the history of acoustics, dealing both with theories of aerial sound and vibrations of solids, was thoroughly explored elsewhere, in introductions to the works of Euler (Euleri Opera Omnia, Series II, volume 11, second part, 1960, and volume 13, 1955) by C. A. Truesdell, who carried on a lively, complementary and complimentary correspondence with Hunt while each was preparing his manuscript.*

Perhaps these omissions from *Origins in Acoustics* acted as one of the barriers to the book's completion. But the source of the difficulties with the manuscript may have been, ironically, acoustics itself. Hunt, like others, had always been both blessed and plagued by the enormous scope of the field of acoustics: blessed, because an acoustician is often involved with issues that touch the lives of individuals in many walks of life; plagued because an acoustician often has his feet in too many camps, without feeling any one of them is home base.

---

*Frederick Hunt had not drafted an acknowledgments page when he died, but notes indicated that he wished especially to thank Clifford A. Truesdell, Professor of Rationale Mechanics of The Johns Hopkins University, and I. Bernard Cohen, Professor of the History of Science, Harvard University, for their invaluable assistance during the preparation of this manuscript.

Hunt was one of the few individuals who could handle this dichotomy. He was both Gordon McKay Professor of Applied Physics and Rumford Professor of Physics at Harvard. He was active in and received the highest awards of both the Audio Engineering Society (most notably for his work on phonograph records and lightweight tone arms) and the Acoustical Society of America (mainly for his work in underwater acoustics and "sonar," an acronym first coined by him). His interest in architectural acoustics—stimulated by his MIT-educated architect wife, Katharine Buckingham—was both theoretical and applied. His influence pervaded the private and public sector alike. He received the Presidential Medal for Merit in 1947 and the Navy's highest civilian award, The Distinguished Public Service Medal, in 1970.

But Hunt will be remembered best as an educator—prodding, urging, cajoling, and questioning his students, teaching and crusading in his field. He held his students to uncompromising standards; and perhaps it is because he held himself to the same standards that his manuscript on the later history of acoustics and another one on physical acoustics still remain on the back burner. As a former student of Ted Hunt, I take pride in having been able to play a part in bringing to publication this book based on his partial manuscript, but I could wish that he had heeded the advice of Huygens, whom he quotes in his text, that "it was better worth while to publish this writing, such as it is, than to let it run the risk, by waiting longer, of remaining lost."

Robert Edmund Apfel

PREFACE

This book started out to be an introductory chapter on the history of sound for a projected textbook devoted to the subject of physical acoustics. I had supposed, naïvely as it turned out, that the assembling of such material would be a straightforward task of referring to the literature on the history of sound. To my dismay, this literature turned out to be unaccountably meager. In my attempt to identify a few quotations and check a few references, I found one footnote leading to another and no satisfying way to stop the chain reaction except to keep on until I ran out of footnotes. If this doesn't qualify as a reductio ad absurdum, it can certainly stand as an illustration of the primitive mathematician's "method of exhaustions." Needless to add, the material soon outgrew the appropriate dimensions of an introductory chapter. The net result of the undertaking is this book.

The lack of a definitive history of the science of sound made it necessary to use, as a guide to the source material, fragmentary historical notes and footnote references appearing in a wide variety of textbooks and treatises on the various topics in physical science that overlap or make peripheral contact with the field of physical acoustics. Before long I woke up to the fact that this pattern itself had a special significance. For many years I have preached to students, or to anyone else who would listen, that acoustics is a "crossroads" subject—that it bears its richest fruits when regarded as a synthesis of other classical disciplines rather than when pursued in textual isolation. The significance I found in the pattern of the source material was that this conception of the subject matter of acoustics also characterized its pattern of historical evolution. Moreover, I believe it is this feature which accounts for the paucity of historical literature in this field; the historical origins in acoustics at any stage of its development are not, for the most part, to be found in the state of acoustics in a preceding era, but in the antecedent history of mathematics, mechanics, heat, and a handful of the other disciplines that contribute in a two-way intellectual exchange with the content of modern acoustics. These convictions about modern acoustics are set forth in some detail in the Introduction.

There are striking differences in the types of relevant source material that pertain to different periods in the development of science. For example, any examination, beyond the most superficial, of the state of science in antiquity confronts the amateur with the intimidating problems of the highly technical field of doxography. In addition to linguistic hazards, which I was ill prepared to cope with, the neophyte is confronted with what I cannot refrain from labeling as a cult of obscurantism. The science of abbreviation has achieved at the hands of these specialists a degree of compactness rivaling the eclecticism of the matrix calculus. These difficulties served to make me even more acutely conscious of my amateur standing as a historian. I can only hope that any gaucheries exposed in part 1 will serve to goad someone better qualified into doing justice to the wealth of extant material. The splendid *Source Book in Greek Science*, by I. E. Drabkin and M. R. Cohen, was an eye-opener for me, its mine of information being valuable not only for the excellent selection of translations, but also for its critical notes and its ancillary guides to the literature dealing with the period. Many of the quotations for which I have given references to the originals appear also in this *Source Book*.

As for the period extending roughly from Boethius to

Leonardo da Vinci, these ages would be a lot darker, scientifically speaking, if it were not for the late G. Sarton's monumental *Introduction to the History of Science.*

The source material dealing with scientific progress in the sixteenth and seventeenth centuries is, if anything, more obscure than for the period of antiquity. The material is old enough, and printing was young enough, that most of the original books and manuscripts are rare and must be properly guarded as library treasures. Besides, enough of the original material *is* still available, and science has marched so rapidly with time, that much of the incentive has been removed for the kind of recensions, translations, and commentaries that grew up around the documentary tradition of antiquity. Subsequent "Collected Works" are relatively less rare, but many of these lose the flavor of the originals.

Keenly mindful of my own grumbling about incomplete, overabbreviated, and incorrect references, I have tried sedulously to avoid these sins. Every reference to the literature has been checked against the original document or publication cited, all but a handful of them being available in the splendid collections of the Widener and Houghton Libraries of Harvard University.

With regard to the spelling of names and places appearing in the text, I have relied chiefly on the *Oxford Classical Dictionary* for the antique period. Elsewhere I have leaned heavily on J. C. Poggendorff's *Biographisch-Literarisches Handwörterbuch.* In documentary references, however, I have followed the orthography of the library catalogue listing, or else that of the title page itself. Dates following proper names are derived, when possible, from the *Oxford Classical Dictionary*, Poggendorff, or G. Sarton's *Introduction to the History of Science.* In quoted material, square brackets enclose explanatory material supplied from context by me or the translator; ordinary parentheses within quotations belong to the material quoted.

F. V. H.

INTRODUCTION

Man lives in an uneasy ocean of air continually agitated by the disturbances called sound waves. Much of our acoustical experience is involuntary, but the production of sounds that please as well as serve has been a proper concern of man ever since he learned to smile.

Sound "is one of the subtilest pieces of nature," Sir Francis Bacon wrote early in the seventeenth century. He added, "so hath the nature of sounds, in some sort, [been inquired into] as far as concerneth music. But the nature of sounds in general hath been superficially observed."[1] The interest in the sounds of music that Bacon noted requires little explanation and no defense, but further inquiry is called for to determine the justification for his charge of superficiality in the study of "sounds in general." In the light of modern understanding, it can be said that nearly all branches of physical science had been only superficially observed in Bacon's time. The seeds of analytical self-consciousness were already there, however, and Bacon's libel against acoustics was eventually discharged through the flowering of a clearer comprehension of the physical nature of sound.

Just as the tides do not rise at the same time on all seacoasts, so the major gains in understanding have not been achieved at the same time on all frontiers of natural science. The underlying causes for this recurrent pattern of stepwise progression are to be found in the essential interrelations among the classical subdivisions of science, no one of which

1

can truly support the pretense of independence. As a consequence, the proper study of historical origins in any branch
of science must be conducted with due and sympathetic regard for these interrelationships. In particular, a just and
proper historical accounting in the field of acoustics makes
unique demands for emphasis on such functional relations—a
circumstance that deserves careful study in its own right.

In preparing a history of any particular science, it is necessary to know at the outset, or else to specify arbitrarily, just
what that science is, what it comprehends, and what distinguishing features set it off from other sciences. Most of the
classical subdivisions of physical science give little trouble in
this connection, and it is easy to define their purview in such
a way that their entire history of development can be presented in a single frame of reference. Thus, when Ernst Mach
opens his historical account of the *Science of Mechanics*[2] by
specifying the subject to be "concerned with the motions and
equilibrium of masses," he invokes concepts that serve equally
well in dealing either with the speculation of the ancients or
with the most sophisticated convictions of the moderns.

For better or worse, it does not appear possible to set up
any such invariant frame of reference for dealing with the
history of acoustics. The relationship between the other
sciences and the science of sound has played an increasingly
prominent role in determining the scope of the subject matter of acoustics, and substantial changes in both have accompanied the gains achieved during two millennia of progress
toward understanding the physical nature of sound itself.
Tracing out the evolutionary course of these changes in scope
and interrelationship must, therefore, be added to the
historian's usual chore of accounting for *who* made *what*
gains *when.*

Prudence normally dictates that historical accounting stop
short of involvement with contemporary developments. To
have accepted this limitation in the present undertaking, however, would have been to ignore the profound changes in the

scope of acoustics that have occurred within the last three decades.* The effect of this expansion of scope, like a mutation in the pattern of evolution in physical science, has been to bring the science of sound into a new estate; and the role that acoustics is beginning to assume in the mid-twentieth century turns out to be so different from its status throughout previous history as to constitute something new and exciting on the acoustical horizon.

The science of sound has never been a branch of learning which could be pursued with maximum profit in isolation, and in modern times it has come to occupy a unique central position in the sensitive area of physical science where there is substantial overlapping of various so-called independent disciplines. This unique position at the crossroads of classical physics is a dominating feature of the "new" physical acoustics, and its implications warrant careful study at the outset, even before its origins are sought in the historical record. Both the uniqueness of this central position for acoustics and the timing of its occupancy need to be explained, if not defended. With regard to timing, the record will indicate—to continue the metaphor—that sound has always been there at the crossroads, but a lot fewer highways led into the intersection during ancient and medieval times. The uniqueness of its central role, like the *why* of its recently expanded scope, is inherent in the physical nature of sound itself; in the circumstances that lead to the sharing of its technical goals with other disciplines; and in the fact that nearly all its subject matter lies in areas where three or more other branches of physical science make common cause.

Casting modern acoustics in a central role, with interdependence as its dominant feature, must not be misunderstood as arrogating to acoustics a dominant role in each of its contributory sciences. On the contrary, the tactical problem

Editor's note: Frederick Hunt wrote these paragraphs in the expectation of completing all four chapters of his history of acoustics.

in acoustics is rather to keep the subject from committing suicide. The hazard of interdependence becoming subservience can be illustrated, in a qualitative way at least, by posing a series of rhetorical questions. Sound waves are produced by the action of moving bodies in contact with air; but is not the motion of material bodies the proper concern of mechanics? Sound waves consist of pressure and velocity disturbances traveling in a compressible fluid; but are not motions and pressure distribution in fluids the proper concerns of hydrodynamics? The condensations and rarefactions that comprise sound waves give rise to regions of higher and lower temperature, between which heat may flow; but are not temperature changes and heat transfer the proper concerns of thermodynamics? And so on. Even when these queries get around to the sense perceptions of hearing, do not physiology and psychology stand ready to intercept? And if the acoustician seeks to cling with desperate finality to the aesthetic appreciation of music, is it not safe to predict that philosophy will also be willing to accept jurisdiction?

The most important conclusion to be drawn from this set of questions is not the obvious one, that each separate technical aspect of an acoustics problem can indeed be parceled out to some other branch of science; it is, rather, that acoustics problems *always* involve such sharing of jurisdiction with at least three other branches. The difference between acoustics and the other sciences in this respect is subtle but not insignificant. Each of the other standard disciplines is primarily concerned with some area of knowledge in which it can claim a substantial degree of self-sufficiency. For example, the idealized problem of a body moving freely in a central force field can be dealt with by mechanics in virtually complete isolation. Yet even under the most ideal circumstances, acoustics must lean on mechanics *and* elasticity *and* thermodynamics as a necessary consequence of the basic definition of sound as wave motion in a deformable material medium. Still other sciences—meteorology or oceanography, electric communications technology, the properties of materials, statistics, psychology—all demand consideration, singly or in groups, depending on what factors are admitted to concern in connection with sound generation, transmission, and reception or perception.

A fair summary of what the foregoing implies is contained in the statement that acoustics is characterized by its reliance on *combinations* of physical principles drawn from other sources; and that the primary task of modern physical acoustics is to effect a fusion of the principles normally adhering to other sciences into a coherent basis for understanding, measuring, controlling, and using the whole gamut of vibrational phenomena in any material medium.

The basic physical nature of the phenomena of acoustics does not become altered merely by a growth in man's understanding of it. By the same token, the feature of interdependence is not made any less an intrinsic attribute of acoustics merely because it was so little noticed by early scientists. Frontiers in science, however, are usually no more than frontiers in the minds of scientists; and it can fairly be said that the growth in appreciation of the fruitfulness of exploiting the interdependence of acoustics and the other sciences has had as much long-range significance as the growth in technical understanding itself.

The characterization of sound as a fusional subject nourished by all the sciences is necessarily a modern concept, since some of the subjects ultimately to be involved in fusion acquired their own independent stature only within the last century and a half. In ancient and medieval times, sound could be studied in splendid musical isolation. The sole access to its phenomena was that afforded by hearing, and this set the subject quite apart from the rest of physical science and restricted its study to the level of superficiality that Bacon duly remarked. The extreme difficulty of making objective observations then served, unfortunately, to sustain this artificial isolation. During the nineteenth century the needed components for profitable fusion started to become available through the maturing of new concepts in heat, electricity, and mechanics; but there was still very little for the appreciation of interdependence to feed on, since measurement, control, and effective use of sound were not yet within reach, at least by modern standards. Much has already been said in other places about the boon of effective instrumentation conferred upon acoustics in the twentieth century by the electronic arts of communication. This was indeed the mutationlike break in the evolution of acoustical science that made it possible for the first time to exploit effectively its expanding areas of overlap with other parts of physical science.

The acoustician is entitled to some concern as to how the word can be spread more widely that his science has something new to contribute in border research areas. "Appreciation of expanded scope" lacks the objectivity of a new result of theory or experiment; and its wide diffusion tends to be inhibited by some of the same factors that helped create it, since intensive specialization seems to follow inexorably on the heels of expansion and growth in science and engineering. As one result, intellectual borrowing across the boundaries between disciplines soon becomes handicapped by specialized terminology and by inevitable preoccupation with special-interest publications. That the problem of interdisciplinary communication is a real one, however, was illustrated not long ago when a distinguished contemporary physicist remarked casually, "I thought all the important problems in sound had been solved, and that this subject had been reduced to a matter of engineering practice." During the ensuing colloquy, part of the basis for his remark was revealed by his startled query, "Oh, do you include vibrations in solids as part of acoustics?" Indeed! It can be added that his tacit recantation of such acoustical heresy can be inferred from his request that his name be withheld.

It has seemed worthwhile to elaborate in this Introduction the distinguishing features of modern physical acoustics, even at risk of making this appear to be an introduction to the future, rather than the past, history of the subject. In thus establishing the point of view that has guided the preparation of these notes, it should also have been made clear that historical origins *in* acoustics have been sought in the record no less avidly than origins *of* acoustics and that the latter have as often as not turned up in fields, and at the hands of scientists, not traditionally associated with acoustics. In short, the foregoing has dwelt on what acoustics has become: now the record must be allowed to tell the story of how and when it got that way.

1 ORIGINS IN OBSERVATION: PYTHAGORAS TO BOETHIUS

The sounds of music and speech are deeply rooted in man's evolutionary past, but the science of sound had its origin in the study of music and vibrating strings by Pythagoras (ca. 570–497 B.C.) during the sixth century B.C. This philosopher of Samos and Crotona, and his master, Thales of Miletus (ca. 640–546 B.C.), were the intellectual pioneers who introduced and established mathematics in the culture of ancient Greece. Pythagoras is primarily remembered now for his espousal of the science of numbers, and this doctrine shaped nearly all inquiries about the nature of sound for the next few centuries. He was also one of the first to insist that precise definitions should form the cornerstone for logical proofs in geometry, although he is better remembered in this field for the unhistorical association of his name with the already well-known theorem about the sums of the squares on the sides of a right triangle. His teacher, Thales, the first of the Seven Wise Men of Greece, had already brought deductive rigor to bear on geometry by introducing the concept of logical proof for abstract propositions, but he, too, is more often remembered in science histories for the unhistorical association of his name with primitive knowledge of the electric properties of amber, the transliterated Greek name for which yields the now ubiquitous root *electron.*

The Ionian school founded by Thales, and the Pythagorean, were the first of the celebrated schools of Greek natural philosophy that were to dominate scientific and intellectual progress for a millennium. The close followers of Pythagoras formed an eclectic group that considered it highly improper to reveal the inner secrets of their philosophy to outsiders, or even to the probationers who attended the lectures Pythagoras gave. This early manifestation of the principle of "security by secrecy" was eventually abandoned; but, unfortunately, it was adhered to long enough to explain in part why no prime accounts of any kind are now available from the hands of the early Pythagoreans. The earliest extant documents that purport to deal with Pythagorean doctrine are fragments attributed to Philolaus of Crotona (fl. fifth century B.C.), a contemporary of Socrates.

The early historians attempted to fill in such gaps in the record, but many of their accounts are necessarily based on no firmer foundation than tradition and legend. Even when special circumstances did not inhibit documentation, many known manuscripts were lost or destroyed, to survive only in translation or commentaries written many years, or even centuries, later. As a consequence, a large element of uncertainty must be conceded in many of the assignments of discoveries or beliefs to specific individuals who flourished in antiquity.

The Arithmetic of Consonance

The contributions of the Pythagorean school to the science of sound were primarily concerned with the science of musical intervals, the branch of musicology sometimes referred to in ancient writings as "canonics," or "harmonics." The musical consonances described as the octave, the fifth, and the fourth were almost certainly known long before Pythagoras; but the success Pythagoras had in identifying these consonances with the ratios of simple whole numbers was not only a major advance in the theory of music but also a source of encouragement and support for the numerological slant of Pythagorean doctrine.

In the early experiments ascribed to Pythagoras, the auditory judgment of consonance was used as a criterion for establishing the corresponding numerical ratios. In due course, however, Pythagoras and his followers lost faith in the evidence of the senses as a criterion of judgment and sought to interpret all phenomena as manifestations of mathematics.[1] To cite the example of a few second- and third-generation Pythagoreans, Heraclitus (ca. 536–470 B.C.) revealed this trend by claiming "the eyes are more exact witnesses than the ears,"[2a] and both "the eyes and ears are bad witnesses for men [if their souls lack understanding]."[2b] Anaxagoras (ca. 499–428 B.C.) explained it more explicitly by saying that "through the weakness of the sense-perceptions, we cannot judge truth."[2c] "Actually," as Philolaus put it, "everything that can be known has a Number; for it is impossible to grasp anything with the mind or to recognize it without this [number]."[2d]

With such a broad scope proposed for the concept of *number*, it is not surprising to find Archytas (428–347 B.C.) describing mathematics as composed of the "related studies" *astronomy, geometry, arithmetic,* and *music.*[2e] The content of these branches of mathematics, according to Pythagorean doctrine, can be inferred from the interpretation of geometry as magnitudes at rest, astronomy as magnitudes in motion, arithmetic as numbers absolute, and music as numbers applied. Plato (ca. 429–347 B.C.) also suggested and rationalized a similar partition of mathematics in his *Republic.*[3] Eventually, after the deification of pure number had diminished somewhat, these four divisions of learning came to be designated as the classic *quadrivium*, which survived into medieval times as the scientific half of a basic program of higher education.

The influence of Pythagorean acoustics research on the development of astronomical concepts, and the measure of Pythagorean faith in the magic of numbers, are both revealed by the numerous attempts to impute musical harmony to the organization of celestial bodies. In the process of asserting the untruth of this idea, Aristotle (384–322 B.C.) gave the

most articulate explanation of the extravagant notion. "Some
thinkers suppose," he wrote, with obvious reference to the
Pythagoreans,

> that the motion of bodies of that [astronomical] size must produce
> a noise, since on our earth the motion of bodies far inferior in size
> and in speed of movement has that effect. Also, when the sun and the
> moon, they say, and all the stars, so great in number and in size, are
> moving with so rapid a motion, how should they not produce a sound
> immensely great? Starting from this argument, and from the observa-
> tion that their speeds, as measured by their distances, are in the same
> ratio as musical concordances, they assert that the sound given forth
> by the circular movement of the stars is a harmony.[4]

Further details about this celestial harmony were supplied
by a later commentator, Alexander of Aphrodisias (fl. early
third century A.D.), who added on behalf of the Pythagoreans, "the sound which they [the planets] make during this
motion being deep in the case of the slower, and high in the
case of the quicker; these sounds then, depending on the ratio
of the distances, are such that their combined effect is harmonious."[5] Aristotle went on to press his attack on this idea
by pointing out that, since "it appears unaccountable that we
should not hear this music, they explain this by saying that
the sound is in our ears from the very moment of birth and is
thus indistinguishable from its contrary silence, since sound
and silence are discriminated by mutual contrast. . . . But, as
we said before, melodious and poetical as the theory is, it cannot be a true account of the facts."[4] It is hardly necessary to
add that the "true account" Aristotle proceeded to offer as an
alternative is not very much easier to swallow than the one he
decried. And in spite of Aristotle, efforts to make metaphysical music with the planets continued to turn up often
throughout the period of antiquity and were to survive in the
serious writings of Kepler nearly two thousand years later.

The most enduring contribution Pythagoras made to acoustical theory was to establish the inverse proportionality between pitch and the length of a vibrating string. The only

accounts of his experiments are found in the later commentaries; and many of these seem regrettably overlaid with legend, since they include descriptions of some experiments that could hardly have been carried out with the indicated results. Among these factitious experiments, one of the most persistent is the legend of the hammers, which has Pythagoras being led to his doctrine of harmonious ratios by the chance observation of consonant sounds produced by a certain set of hammers being used in a metal-working shop. According to the legend,[6] Pythagoras found that the two hammers whose sounds were an octave apart had weights in the ratio of 2:1, that one of these weighed four-thirds as much as another hammer with which it gave the consonance of a fourth, and so on. The physical factors that determine the dominant pitch of the sound produced by a hammer hitting an anvil are, to be sure, far from simple; but it can be stated with considerable confidence that the alleged results ascribed to Pythagoras were either spurious or coincidental. Similar invalid conclusions were also ascribed to Pythagoras in connection with the change of pitch that occurs when a vibrating string is subjected to the changes of tension produced by different hanging weights, and in connection with the sounds of "musical glasses" and the tones generated by striking various partially filled vessels.

The Pythagorean tenet of secrecy undoubtedly contributed to the durability of these legends, since it forced historians to rely on the secondary source material of earlier commentaries. But while this circumstance lends uncertainty to the chronology and authorship of new ideas, it does not in the least depreciate the value of such historical commentaries as a catalogue of prevailing notions, both valid and invalid. Thus, when Boethius (A.D. 480–524) tells of Pythagoras inventing the monochord *in order to carry out experiments* dealing with the relation of pitch to the length of a vibrating string (see n. 6), it is perhaps of less importance that his account be accurate than for him to reveal, for his own time if not for that of Pythagoras, the conception of man-made apparatus used deliberately for a scientific experiment.

While the conclusions drawn by Pythagoras, in the sixth century B.C., about the inverse proportionality of pitch and string length were entirely valid, there is some reason to doubt that Pythagoras himself had grasped the concept of frequency or that he understood its relation to pitch. However, two of the later Pythagoreans, Archytas, the mathematician of Tarentum mentioned above, and Eudoxus of Cnidas (ca. 408–355 B.C.), a distinguished mathematician and astronomer, do seem to have achieved at least a qualitative understanding of this relation. For example, Theon of Smyrna (fl. ca. A.D. 115–40) wrote that "the schools of Eudoxus and of Archytas held that the relations of consonance could be expressed by numbers. They understood also that these ratios represented the motions, a rapid motion responding with a shrill tone, because it strikes and goes through the air more rapidly and continuously, and a slow motion answering with a deep tone, because it is more sluggish."[7]

The line of testimony is a little more direct in a rediscovered fragment ascribed to Archytas, in which he said of the sounds that "impinge on the perception, those which reach us quickly and powerfully from the source of sound seem high-pitched, while those which reach us slowly and feebly seem low-pitched. For if one takes a rod and strikes an object slowly and feebly, he will produce a low note with the blow, but if he strikes quickly and powerfully, a high note. . . . Clearly swift motion produces a high-pitched sound, slow motion a low-pitched sound."

Only a vague concept of frequency is revealed by these extracts; and there is obvious confusion between the speed of the blow producing the sound and the speed of sound transmission, and perhaps also between the concepts of intensity and pitch. There is no uncertainty, however, about Archytas' conviction "that sound is impossible unless there occurs a striking of objects against one another"; and he concludes his argument with a summary claim "that high notes are in swift motion, low notes in slow motion, has become clear to us from many examples."[8]

The mistaken idea that the speed of sound varies with the pitch had not been completely dispelled even in Aristotle's time, although Aristotle seems to be less than certain on this point when he says, "some of the writers who treat of concords assert that the sounds combined in these do not reach us simultaneously, but only appear to do so."[9] Instead of coming to grips with this physical question, however, Aristotle diverts himself to a discussion of simultaneity in sense perception. The proper physical deduction was made, however, by Theophrastus of Eresus (372–288 B.C.), the naturalist who became Aristotle's successor as head of the Peripatetic school. Theophrastus reasoned that "the higher note would not differ in speed [from the lower], for if it did it would lay hold on the hearing sooner, so that there would not be concord. If there is concord, both notes are of the same speed."[10] Precisely this same line of reasoning was to be invoked by Jean Henri Hassenfratz in 1805, when he demonstrated by direct experiment that the concordant sounds of two bells struck simultaneously could be heard at a distance of half a kilometer without modification of the consonance.[11]

A modern inheritance from the Pythagorean doctrine of simple numbers is the musical scale now identified as Pythagorean intonation. Theon of Smyrna assigns directly to Pythagoras the identification of the principal interval ratios;[12] but it was the prevailing custom to assign all the discoveries of a "school" to its master (even in those days!) and Theon's terminology suggests that he may have been drawing on Euclid or Aristotle. Others credit Pythagoras with adding a string to the lyre and find in this an origin for the Pythagorean "scale." However, what seems to be the earliest, and one of the clearest, descriptions of the elements of the Pythagorean scale is contained in a fragment ascribed to Philolaus, which runs, the content of the octave is the major fourth and the major fifth; the fifth is greater than the fourth by a whole tone; for from the highest string [i.e., the longest string (of the lyre) and lowest note] to the middle is a fourth, and from the middle to the lowest string [i.e., the shortest string and highest note] is a fifth. Between the middle and third strings is a tone. The major fourth has the ratio 4:3, the fifth 3:2, and the octave 2:1. Thus the octave consists of five whole tones and two hemitones, the fifth consists of three tones and a hemitone, and the fourth consists of two tones and a hemitone.[13]

The sequence of the whole tones and hemitones in the upper and lower portions of the octave is not defined uniquely by this passage, but the data given do establish the Pythagorean whole-tone interval as 9:8, and the hemitone interval as 256:243. Among the various permutations of order allowed by Philolaus' specification, the one which goes, in ascending sequence, whole-whole-hemi-whole-whole-whole-hemi is the one now referred to as Pythagorean intonation or "Lydian mode." Ringing changes on the musical scale was a confirmed delight of many ancient musicians, and the questions of modality are aptly termed one of the thorniest in the history of music. No enduring modifications of the Pythagorean scale were made, however, until Aristoxenus proposed a subdivision of the whole-tone interval and until the two minor whole tones (i.e., two intervals of 10:9) were introduced as elements of the diatonic scale by Didymus and Ptolemy (see below, p. 30).

Several musical topics are included in the natural science catechism, the *Problemata*. It is customary to refer to the author of this interesting compilation as "Aristotle," because he is believed to have written a work of this kind; but the consensus of opinion is that Aristotle himself did not write it. It is conceded, however, that the *Problemata*, except where it contradicts itself, reflects accurately the views of the Peripatetic school. The following musical question illustrates this by exhibiting a rather severe example of Aristotelian logic: "Why are a double fifth and a double fourth not concordant, whereas a double octave is?" The question is answered by reasoning that "in a series of three numbers in a ratio [each to the next] of 3:2 or 4:3, the two extreme numbers will have no ratio to one another; for neither will they be in a super-particular ratio [i.e., of the form (n+1):n], nor will one be a multiple of the other [i.e., of the form n:1]. . . . So, since a concord is a compound of sounds which are in a proper ratio to one another, . . . the sounds constituting the double octave would give a concord while the other would not, for the reasons given."[14] It is noteworthy that this problem is considered to be suitably resolved merely by showing that the discordant tones "have no ratio to one another."

The great geometer Euclid of Alexandria (330–275 B.C.) probably needs to share only with Archimedes (ca. 287–212 B.C.) and Apollonius of Perga (ca. 262–190 B.C.) the distinction of being one of the greatest mathematicians of antiquity. However, Euclid's treatment of musical intervals is not very much more penetrating than that revealed in the *Problemata*, although Euclid presents the doctrine of simple numbers in a somewhat more orderly syllogism. This begins:

> If there were complete rest and immobility, there would be complete silence. . . . All sounds result from some blow. . . . Some sounds are higher pitched, being composed of more frequent and more numerous motions, . . . Sounds lower pitched than what is required reach the required pitch . . . by an increase in the amount of motion. Therefore sounds must be said to consist of parts, since they reach their proper pitch by addition or subtraction. Now all things that consist of parts may be spoken of as in a numerical ratio to one another. And this must be the case also for sounds, which are thus said to stand in a numerical ratio to one another. . . . Thus it is reasonable to say that since consonant sounds join to produce a single blend of tone, they belong to numbers whose relation may be expressed . . . [as] being in multiplicate or superparticular ratio.[15]

This passage is taken from the *Sectio Canonis*, which is probably the "elements of music" ascribed to Euclid by most of his commentators.[16] The opening phrases of this passage also indicate that the vagueness of Archytas' conception of frequency and its relation to pitch had been at least partially cleared up during the intervening century.

The Philosophical Anatomy of Sound and Hearing

In his *Apology*, Plato has Socrates (469–399 B.C.) characterize himself as a "gadfly . . . fastened on the city . . . arousing, and urging, and reproaching each one"[17]—an epitome of what continues to be one of the useful functions of an educator. Socrates rejected the Pythagorean devotion to number as the true reality on the ground that it is a delusion to think that true knowledge is attainable. He turned instead to the development of a doctrine of ethics based on the autonomy of the individual intellect. The Socratic method of instruction by means of question and dialogue made him the first exemplar of the educational principle that students may be helped to learn but cannot be taught.

Socrates made no pretense toward a scientific theory of morality and can hardly be said to have founded a school of philosophy of his own, owing to his disclaimer of any unique Socratic doctrine. However, four different schools of philosophy did spring up at the hands of his pupils and their followers, each responding to and developing different facets of the "teaching" of Socrates. The hedonistic Epicureans seized upon the Socratic precept of promoting the well-being of the individual, while the Skeptics sought to elevate Socrates' agnosticism. In sharp contrast with the Epicurean viewpoint, the Stoics took as doctrine the austere disdain with which Socrates had regarded his own personal well-being.

The best known of these schools, however, was the Academic, founded by Plato, in whose philosophy was merged much of the best of both Pythagorean and Socratic principles. To the late Pythagorean doctrine that knowledge is to be sought through reason and not in the evidence of the senses, Plato added the Socratic concern for human improvement; and ultimately he went far beyond Socrates in asserting his faith in the power of mind to attain true knowledge. By his pursuit of exactitude in reason, Plato was led, especially in the later years of his long life, to make substantial contributions toward understanding the nature of generalization and toward a systemization of the logical methods that could properly be used as a basis for mathematical analysis.

Insofar as the specific contributions of Plato to the science of sound are concerned, his name might appropriately be linked with that of his pupil Aristotle in a paraphrase of the nineteenth-century epigram that claimed the greatest scientific discovery of Sir Humphry Davy to be Faraday. Since this is a way of saying that Plato did not contribute very much to the science of sound, it should be noticed that he does find occasion, in his *Republic*, to deprecate the empiricists in musical science—"those gentlemen who tease and torture the strings and rack them on the pegs of the instrument . . . setting their ears before their understanding." However, Plato does not quite align himself with the Pythagoreans either, "for they too are in error, like the astronomers; they investigate the numbers of the harmonics which are heard, but . . . they never reach the natural harmonies of number, or reflect why some numbers are harmonious and others not." "That," said Plato, "is a thing of more than mortal knowledge."[18]

Elsewhere, Plato had somewhat more to say about hearing, and "the causes in which it originates." On this important subject he wrote:

We may in general assume sound to be a blow which passes through the ears, and is transmitted by means of the air, the brains and the blood, to the soul; and that hearing is the vibration of this blow, which begins in the head and ends in the region of the liver. [!] The sound which moves swiftly is acute, and the sound which moves slowly is grave; and that which is regular is equable and smooth, and the reverse is harsh. A great body of sound is loud, and a small body of sound the reverse.[19]

However one may react to the startling suggestion that hearing is a motion localized near the liver, it is clear that no

very high level of sophistication prevailed during the pre-
Christian era with regard to the anatomy of hearing. In spite
of an impressive growth of empirical knowledge of the skeleton and the circulatory systems of the body, the minuteness
and delicacy of the hearing mechanism guarded its secrets
from all but the most superficial inquiry. Theophrastus, who
contributed to these inquiries, summarized the beliefs of his
predecessors in his treatise *On the Senses*. To cite a few of
these, Alcmaeon of Crotona (fl. ca. 500 B.C.) is quoted as
saying that "hearing is by means of the ears, because within
them is an empty space, and this empty space resounds."[20a]
Note, however, that the term "empty space" was often used
by these early Greek scientists to denote space containing
nothing but air. This general idea was sharpened up a century later by the Sicilian, Empedocles (ca. 493 B.C.). After
characterizing the external organ of hearing as "a Fleshy off-
shoot," he went on to say "that hearing results . . . whenever
the air, set in motion by a voice, resounds within. For the
organ of hearing . . . acts as the bell of a trumpet, ringing
with like sounds."[20b]

In his account, Theophrastus then had Anaxagoras meet
the central issue with naked simplicity: "hearing," he said,
"depends upon the penetration of sound to the brain."[20c] The
Roman poet Lucretius (94–55 B.C.) dealt almost as summari-
ly with hearing when he wrote that "every sound and voice
is heard, when creeping into the ears they have struck with
their body upon the sense."[21] Anaxagoras, however, added
an unsettling remark about the access for "penetration to the
brain"—"sense perception varies in general with the size," he
wrote. "Large animals hear loud sounds and sounds far away,
and the more minute sounds escape them; while small animals hear sounds that are minute and close at hand" (see n.
20c).

With regard to the mechanism of hearing, Theophrastus
found himself in substantial accord with his master Aristotle,
who had described its action by saying, "The organ of hearing

is physically united with air, and because it is *in* air, the air inside is moved concurrently with the air outside."[22] While this conclusion is not impressively weighty, it does supply a mechanism for Empedocles' "resounding within," and it has some further interest as an anticipatory suggestion of the transmission of sound by virtue of the action of air on air.

The Physical Nature of Sound

As the philosophic heir of Plato, Aristotle embraced much of his doctrine, but he succeeded in avoiding some of the obscurity of Plato's mysticism by devoting himself to the transition from a philosophy of universals to a philosophy of individual substances. Aristotle probably deserves to be called the first mathematical physicist, since he was deeply concerned with the whole range of natural philosophy and with the use of mathematical reasoning as a tool for examining nature. If some of his conclusions now seem fanciful, this should not belittle the sound logic he often used in reaching them nor his mastery of some physical principles that were not widely understood in his time. For example, he had a clear conception of the force parallelogram and the rules of proportionality for simple levers, although it is doubtful that he should be credited with the discovery of either of these. He also understood the optical principle of equal angles of incidence and reflection, although this too has now been established as probably pre-Aristotelian.[23] However, the Aristotelian author of the *Problemata* does seem to have been the first to point out that the same law of reflection operates for sound.[24]

The relative velocity of transmission of light and sound periodically commanded the attention of philosophers and scientists until nearly the middle of the eighteenth century. In this connection, Aristotle served up a nice mixture of the true and false when he wrote that lightning "comes into existence after the collision and the [resulting] thunder, though we see it earlier because sight is quicker than hearing."[25] Perhaps because it was Aristotle who said it, the inverted notion that thunder causes the lightning was to persist for centuries. There is also a suggestion here that perhaps light and sound both proceed from their sense organs to the object, although elsewhere Aristotle avoids this inversion with regard to sound and hearing.

Two other historians of antiquity expressed proper conclusions concerning these relative velocities. Pliny the Elder (A.D. 23–79), the encyclopedic recorder of the state of natural history in his time, observed that, "It is certain that when thunder and lightning occur simultaneously, the flash is seen before the thunderclap is heard (this not being surprising, as light travels more swiftly than sound)."[26] Perhaps the most remarkable feature of this statement lies in its untroubled use of the notion that light "travels" with a speed which is not so instantaneous but that it can be conceived of as comparable with the speed of sound. Aristotle had wrestled with this problem earlier, in the course of combating an idea advanced by Democritus, but he bogged down on the question of how light could exist, invisible in space, during a finite time-interval between its emission and its perception. Nearly a century before Pliny wrote his *Natural History*, Lucretius seems to have had these ideas well straightened out, and to have expressed them so felicitously as to deserve quotation in Johnson's metrical translation:

> We see the lightning ere the thunder hear;
> For quicker comes the impulse to the eye
> Than to the ear; as plain to be perceived
> By watch of woodman laboring with axe,
> To fell the pride of some wide-spreading tree;
> The falling blows are seen before the sound
> Comes to the ear; so blinding lightnings come
> Before the thunder, though from self-same cause
> Springs thunder and its fleet, vaunt courier, light.[27]

The fundamental idea expressed by Archytas—that sound is always produced by the "striking of bodies on one another"—

was paraphrased in one way or another and repeated by almost every writer of ancient and medieval times who considered the generation of sound. Aristotle's version of this idea is set forth most clearly in his essay *On the Soul*, where he writes: "What is required for the production of sound is an impact of two solids against one another and against the air. The latter condition is satisfied when the air impinged upon does not retreat before the blow, i.e., is not dissipated by it. That is why it must be struck with a sudden sharp blow, if it is to sound—the movement of the whip must outrun the dispersion of the air." This passage contains the germ of an idea that was not to be fully appreciated until Stokes dealt with the matter in the nineteenth century, the idea that the medium must not merely "retreat before" (or circulate around) the sounding body but must be compressed by it. In the same section of his essay, Aristotle explains that "an echo occurs, when . . . the air originally struck by the impinging body and set in movement by it rebounds . . . like a ball from a wall." But Aristotle is writing here of the Soul and so must add the haunting thought, "Not every sound made by an animal is voice . . . what produces the impact must have soul in it and must be accompanied by an act of imagination, for voice is a sound *with a meaning.*"[28] Perhaps this *is* one step beyond Democritus, for he had been content to say merely that "Speech is the shadow of action."[29]

The analogy between the expanding pattern of ripples on a pool of water and the propagation of sound waves also achieved wide currency in ancient times. This analogy has, in fact, been used (and abused) almost as generously as the nineteenth-century analogy between electric current and the flow of water in a pipe. The imagery seems to have been conjured up first by the Stoic philosopher Chrysippus (ca. 280–207 B.C.), to whom the following passage is attributed by the biographer Diogenes Laertius (fl. first half of third century A.D.): "Hearing occurs when the air between that which sounds and that which hears is struck, thus undulating spherically and falling upon the ears, as the water in a reservoir undulates in circles from a stone thrown into it."[30]

Marcus Vitruvius Pollio (last century B.C.), the Roman architect usually referred to as Vitruvius, extended the ripple analogy when he wrote:

> The voice is a flowing breath, made sensible to the organ of hearing by the movements it produces in the air. It is propagated in infinite numbers of circular zones, exactly as when a stone is thrown into a pool of standing water. . . . Conformable to the very same law, the voice also generates circular motions; but with this distinction, that in water the circles remaining upon the surface, are propagated horizontally only, while the voice is propagated both horizontally and vertically.[31]

In the overall conception of sound transmission that is revealed by these various extracts, there is a curious lack of emphasis on the basic idea of compression in the medium. Aristotle's oblique reference to it has already been mentioned, but the Roman moralist Seneca (ca. 5 B.C.–A.D. 65) raised the issue more specifically in his *Natural Questions*, by asking: "what is the voice save tension of the air moulded by a stroke of the tongue so as to become audible? . . . What song can be sung without tension of the breath? Or, take horns and trumpets, or those organs that . . . can produce a greater volume of sound than the mouth is capable of doing: is it not through atmospheric tension that they display their functions?"[32]

The antecedents of those ideas are probably to be found in the curious notions held by the early atomists about the various "shapes" that could be impressed on the air. Thus Democritus held "that the air is broken into bodies of similar shapes and rolls about with the fragments of the sound."[33] Some, if not all, of the followers of Aristotle also fell for this idea, for a question is raised (twice) in the *Problemata* about "the voice, which is air that has taken a certain form and is carried along."[34] The short reference to hearing from Lucretius quoted above shows that he, too, believed sound to have a corporeal form. He gave further expression to this notion by adding later, "When therefore we press out these voices from the inmost parts of our body, and send them forth straight through the mouth, the quickly-moving tongue, cunning fashioner of words, joints and moulds the sounds, and the shaping of the lips does its part in giving them form."[35]

It is almost needless to add that the idea of a shape impressed on the air came under vigorous attack. Theophrastus, in seeking to demolish the view held by Democritus, posed the question "how [can] a few fragments of wind completely fill a theatre containing ten thousand men?" (see n. 33). As an aside, it can be remarked that modern sound engineers are often nearly as embarrassed by this question as Theophrastus hoped the followers of Democritus would be. Theophrastus pressed his attack further[36] on the grounds that the "imprints" could be perceived not only by the listener but also by the speaker, who would always hear an echo; and that the situation would be hopelessly confused when the imprints from several sources must occupy the same space. Thus he finally dismissed the whole notion as extravagant. A more constructive criticism of the concept of imprints was offered by Epicurus (341–270 B.C.) of Samos, a pupil of Plato's who embraced atomism and founded the school of philosophy that bore his name. In a long letter to Herodotus, he observed that, "we must not suppose that the air itself is moulded into shape by the voice emitted . . . for it is very far from being the case that the air is acted upon by it in this way. The blow which is struck in us when we utter a sound causes such a displacement of the particles as serves to produce a current resembling breath, and this displacement gives rise to the sensation of hearing."[37]

By far the most enlightened criticism of the doctrine of imprints, and perhaps one of the most prescient passages in the ancient documents dealing with sound, is contained in *De Audibilibus*, a fragmentary manuscript of uncertain date and authorship. Its opening paragraph presents an admirably compact summary of the physics of sound generation and transmission, in these words:

> All voices and in fact all sounds arise either from bodies falling on bodies, or from air falling on bodies; it is not due to the air taking on a shape as some think, but to it being moved in the same way as bodies, by contraction, expansion and compression, and also by knocking together owing to the striking of the breath and by musical strings. For when the breath that falls on it strikes the air with successive blows, the air is at once driven forcibly on, thrusting forward in like manner the adjoining air, so that the sound travels unaltered in quality as far as the disturbance of the air manages to reach. For, though the disturbance originates at a particular point, yet its force is dispersed over an extending area, like breezes which blow from rivers or from the land."[38]

The authorship of this remarkable document is conventionally assigned to Aristotle, and the manuscript is classified among the "minor works" in the "Aristotelian corpus"; but just as in the case of the *Problemata*, it is generally conceded that Aristotle did not write it. So far as the evidence of internal consistency is concerned, the case is somewhat stronger against the *Problemata*. For example, the statement that "the voice is not audible through water" can be contrasted with at least two statements of the contrary view appearing in other Aristotelian writings of more certain tradition.[39] Another contrast is afforded by the espousal of the doctrine of imprints in *Problemata* (see n. 34) and its denial in *De Audibilibus;* but while this supports the argument that their authors were not the same, it does not provide positive evidence for assigning the authorship of either. However, since Theophrastus does not mention Aristotle in his attack on Democritus' theory of imprints, it seems likely that the *Problemata* must at least postdate Theophrastus.

On the positive side of the authorship question, some historians have ascribed *De Audibilibus* to Heraclides of Heraclea on the Pontus, also called Ponticus (ca. 390–310 B.C.); but it seems most likely to have been the work of Straton of Lampsacus (ca. 340–269 B.C.). Straton, sometimes called "the physicist," was a pupil of Theophrastus and succeeded him as leader of the Peripatetic school on the latter's death in 287 B.C. He sought to combine in his natural philosophy[40] the analytical viewpoint of Aristotle and the atomistic conceptions of Democritus—a difficult merger to bring about, but one that might well represent just the kind of mixed background required for so understanding a description of the propagation of a sound wave by the expanding action of air on air. For, in spite of the animadversions of Theophrastus, Democritus was close to the truth when he spoke of sound in connection with hearing. "Once the commotion has been started within," he wrote, "it is 'sent broadcast' by reason of its velocity; for sound arises as the air is being condensed."[41]

Many of these conceptions and ideas that arose during the ascendant period of Greek science were collected and summarized by Boethius, who was one of the latest of the ancient commentators. After defining sound as a "blow upon air, the effect persisting undissolved until the hearing is reached," he adds in a later section:

> Let us now speak of the method of hearing. In the case of sounds something of the same sort takes place as when a stone is thrown out and falls into a pool or other calm water. . . . In the same way, then, when air is struck and produces a sound, it impels other air next to it and in a certain way sets a rounded wave of air in motion, and is thus dispersed and strikes simultaneously the hearing of all who are standing around. And the sound is less clear to one who stands further away since the wave of impelled air which comes to him is weaker.[42]

Quite apart from questions of authorship, which can only affect the *dating* of such writings as *De Audibilibus*, it is humbling to consider how remarkably little modernization of the language of these ancient writings is required in order to qualify them still to serve as admirable elementary descriptions of the physical mechanism of sound generation and propagation.

It is worth remarking that the doctrine of number reached its peak of acceptance, as a part of Plato's quest for universal absolutes, at about the same time when the Parthenon and the Erechtheum were being placed as crowns on the Acropolis in Athens and while Praxiteles was creating his immortal sculptures. This was Greece's Golden Age, to which Western culture will always be indebted for a classic heritage of art and literature. It might be difficult to establish with any certainty the interaction between Plato's philosophy and the contemporary achievements in sculpture and architecture, but there is at least some refreshment in associating the clean simplicity of the artistic monuments of the Golden Age with Plato's devotion to universal forms.

Temperament and the Structure of Musical Theory

In the founding of a science of music, Pythagoras had relied on a dualistic interplay between numerical relations and auditory sensations. It was pointed out above, however, that the followers of Pythagoras, if not Pythagoras himself, soon abandoned this dualism; and no shift from their viewpoint was involved in Plato's insistence that canonics was simply a branch of mathematics. In their turn, neither Aristotle nor Euclid did very much to alter the numerological status of musical science; but one of Aristotle's pupils, Aristoxenus (fl. ca. 320 B.C.) cast himself in the role of a musical reactionary.

Aristoxenus began, as Pythagoras had, by urging that both auditory judgment and numerical ratios should be considered in musical theory, announcing his position by saying, "our method rests in the last resort on an appeal to the two faculties of hearing and intellect." However, like many another reactionary, Aristoxenus found it difficult to maintain the "golden mean" position advocated by Plato and Aristotle and glorified by Horace (*Odes* 2. 10). The trend of his musical science toward emphasis on subjective judgment is revealed in his treatise on *Harmonics*, where he concludes his introductory account by saying, "every proposition . . . must be such as to be accepted by the sense perception as one of the primary truths of Harmonic science . . . lest . . . we let ourselves be dragged outside the proper track of our science by beginning with sound in general regarded as air-vibration."[43] The physicist may not welcome having his science so explicitly excluded, but at least he cannot say he wasn't warned! This treatise, incidentally, is one of the oldest on musical theory of which substantial portions are still extant, and its continued availability in both ancient and medieval times permitted it to exert a profound influence on the evolution of musicology.

The principal contribution Aristoxenus made to musical theory was to introduce two fractional tones, each smaller than the Pythagorean hemitone, as building blocks for the construction of musical scales in different modes. The one-third tone was called the "smallest chromatic diesis," the quarter-tone the "smallest enharmonic diesis," and no smaller tone intervals than these were allowed as ingredients of melody. Aristoxenus then proposed to use addition to combine these tone intervals in scale construction, rejecting as "extraneous reasoning" his predecessors' practice of fixing the height or depth of pitch by means of "numerical ratios and relative rates of vibration." Of course, as Boutroux has pointed out, the method of adjusting intervals by addition that Aristoxenus proposed is approximately equivalent to computing the usual Pythagorean frequency ratios by means of logarithms.[44] As consequence, his departure from precedent was not as profound as he had intended. Boutroux also points out that the "theoretical" scale to which Aristoxenus leads himself is actually a closer approximation to the modern tempered scale than he is usually given credit for achieving.

Ptolemy (Claudius Ptolemaeus) of Alexandria (A.D. 70–147), the astronomer whose geocentric outlook flourished until Copernicus spoke out in 1543, was the next scientist of antiquity who undertook to unite auditory sensation and numerical relations in the interest of musical science. As a mathematician and experimentalist of some distinction, Ptolemy could not condone the empiricism of Aristoxenus any more than a theoretical physicist can reconcile himself to the unexplained results of the experimentalist. At the same time, Ptolemy was quite aware of the existence of limitations adhering to the Pythagorean scale. Claudius Didymus (first century A.D.) had also recognized this defect and had sought to correct it by introducing a minor whole tone of ratio 10:9. Since the writings of Didymus have all been lost, his work is known now only through the accounts of it given by Ptolemy,[45] who seems to have preserved it largely for the sake of controverting it. Didymus had elected to place the minor tone as the first (ascending) interval of the scale, whereas Ptolemy made it the *second* interval. Including the minor tone in either position served to remove the excess of the major third, the outstanding flaw in the Pythagorean scale. Moreover, for either sequence of major and minor whole tones, the accuracy of the major fourth and fifth could also be preserved by shortening the Pythagorean hemitone ratio from 256:243 to the diatonic semitone ratio of 16:15. The net effect of these changes was to lower the fourth note of the scale slightly, but only by the ratio of 80:81, a very small interval that is still known as the "comma of Didymus."

The very real advantages of Ptolemy's placement of the minor-tone interval only appear when the scale is examined with respect to the consonance of intervals erected on different initial notes, or degrees, of the scale. This aspect of the problem was not fully analyzed, however, until Gioseffo Zarlino (1519–90) undertook it nearly fifteen centuries later.[46] Nevertheless, Ptolemy, with his confirmed mathematical predilections, could not pass up this opportunity to experiment with the various scales, or modes—experiments that involved cataloging all the permutations of the placements of the major and minor whole-tone and semitone intervals. The indiscriminate nature of much of his "wanton sport-

ing with the scale" led the eighteenth-century music historian
Charles Burney to characterize Ptolemy as "possessed with an
unbounded rage for constructing new scales."[47] The musical
scale now referred to as "just intonation," or as the true or
natural diatonic scale, was included as the *intense diatonic*
among the many scales that Ptolemy proposed, and it is
probably to be counted as his most enduring legacy to music.

The problems of modality and musical scale were of much
more than theoretical interest, since there was great demand
for a practical solution that would make it possible for fretted
string instruments or woodwinds to be played in more than
one or two keys without retuning. This practical problem was,
in essence, the major one to which Aristoxenus had addressed
himself, and his near success in solving it might have been
even more influential if the issue had not been clouded by
Ptolemy's "wanton confusion."

The first modern to appreciate that the solution of this
problem lay in a suitable "temperament" was Bartholomeo
Ramos de Pareia (ca. 1440–1521), a Spanish musician known
through his lectures at Bologna. His tract on *Musica Practica*[48]
attracted less attention than did the more complete analysis
published later by Zarlino, to whom credit is usually assigned
for the proposal (see n. 46) of the equally tempered scale.
The growth of interest in keyed string instruments during the
fifteenth and sixteenth centuries made the adoption of some
sort of compromise tuning necessary; and the sufficiency of
Zarlino's proposal for compositions written in any key was
dramatically demonstrated in 1722 when Bach brought forward the first of his two sets of preludes for *Das wohl-
temperierte Clavier.*

As a compromise, the equally tempered scale is a remarkably good one. None of its intervals, save the octave, is exactly perfect, but neither does any depart from its "proper" size
by an amount that is deemed objectionable, or even notice-
able except under carefully contrived listening conditions.[49]
This triumph of the counsel of moderation would probably

have delighted Aristoxenus, although the incommensurate size of the equal half-tone intervals could have given little comfort to Ptolemy or the earlier doctrinaires of simple numbers. But of course this stage of evolution in music came centuries later. As Burney complained, musical scales were little more than sporting exercises in mathematical calculation for Ptolemy: his principal concern was with the harmony of the celestial spheres that rotated in all their splendor around his stationary earth.

Architectural and Applied Acoustics

There is remarkably little evidence in the antique record of any awareness—let alone understanding—of the problems and phenomena now designated as "architectural" acoustics. A few observations of that period can, however, be interpreted with greater appreciation in the light of modern concepts.

To pick one example of traditional distinction, consider the Old Testament description of the curtains of goats' hair that were to be hung in the tabernacle. According to the detailed specifications laid down in Exodus Book 26, their length (measured horizontally) exceeded that of the perimeter in such a way that they would necessarily hang in generous folds. While it might be impious to ascribe an acoustical motivation to these dimensional requirements, the interior treatment called for is strikingly similar to that used in early radio broadcasting studios where reverberation was controlled by hangings of monk's cloth.

There is a more direct reference to the acoustical effect of surface absorbents in the Aristotelian *Problemata*, where it is asked: "Why is it that when the orchestra of a theater is spread with straw, the chorus makes less sound? Is it because, owing to the unevenness of the surface, the voice does not find the ground smooth when it strikes upon it and is therefore less uniform, and so is less in bulk, because it is not continuous?"[50a] Alas, the promise of significance in the opening question seems to peter out in incoherence. There is similar deterioration in the continuity following the intriguing question: "Why are newly plastered houses more resonant? Is it because their smoothness gives greater facility for refraction [= reflection?]? . . . One must, however, take a house which is already dry and not one which is still quite wet; for damp clay gives no refraction of sound."[50b] Finally, one can find the observational basis for the logarithmic nature of sensory response in the question, "Why is it that when one person makes a sound and a number of persons make the same sound simultaneously, the sound produced is not equal [proportional?] nor does it reach correspondingly farther?"[50c] Again, alas, the suggested answers yield little but disillusionment.

Lucretius, writing in the last century of the pre-Christian era, reveals a clear concept of reverberation. By glossing over his retention of the idea of shapes impressed on the air, and by telescoping his discussion in paraphrase, his vivid description of this phenomenon can be set forth as follows:

> When there is no long race for . . . voice to run from start to finish, each of the words . . . must necessarily be plainly heard . . . but if the intervening space is longer than it should be, the words . . . must be confused . . . perceived yet not distinguishable in meaning, . . . so confused must be the voice when it arrives, so hampered. . . . One voice . . . disperses suddenly into many voices . . . some scattered abroad without effect into the air: some dashed upon solid places and then thrown back, . . . deluding with the image of a word. . . . In solitary places the very rocks give back the words . . . so does hill to hill buffet the words and repeat the reverberation. Voices distributed . . . beget even as a spark of fire is often wont to scatter itself into fires of its own. Therefore the whole place is filled with voices, . . . all around boils and stirs with sound. . . . All [light] images tend straight forwards when once they are sped; wherefore no one can see beyond a wall although he can hear voices through it. And yet even the voice itself in passing the wall of a house is blunted and confused when it penetrates the ear, and we seem to hear sounds rather than words.[51]

Anyone who has tried to listen to the unintelligible jumble of conversation transmitted through closed doors or building

partitions—and who hasn't?—will appreciate the accuracy of
Lucretius' concluding observation. One can also find recorded
here the phenomenon of sound diffraction, by which the
voice is heard "beyond the wall," another aspect of sound
transmission about which relevant but unanswered questions
had been raised in the *Problemata.*[50d]

The most significant contributions to architectural acoustics, however, were made by the Roman architect Vitruvius.
As an architect, Vitruvius should probably be assigned to the
do-as-I-say-not-as-I-do school, since the only surviving monu-
ment that can be ascribed to him is a long treatise on architecture. The ten Books of his *De Architectura* did survive,
however, and provided an important foundation for the theory and practice of Renaissance architecture. Of more present
interest, Vitruvius included in this treatise a remarkably
understanding analysis of theater acoustics. He wrote as follows:

> we must choose a site in which the voice may fall smoothly, and not
> be returned by reflection so as to convey an indistinct meaning to the
> ear. For there are some places which naturally hinder the passage of
> the voice, . . . those (dissonant) places in which the voice, when first
> it rises upwards, strikes against solid bodies above, and is reflected,
> interfering as it settles down with the rise of the following utterance
> . . . those (circumsonant) in which the voice moves around, is then
> collected in the middle where it dissolves without the case-endings
> being heard, and dies away in sounds of indistinct meaning . . . those
> (resonant) in which the words, striking against a solid body, give rise
> to echoes and make the case-endings sound double . . . those (con-
> sonant) in which the voice reinforced from below rises with greater
> fullness, and reaches the ear with clear and eloquent accents. Thus if
> careful observation is exercised in the choice of sites, such skill will
> be rewarded by the improved effect of the actor's voices. . . . Who-
> ever uses these rules, will be successful in building theatres.[52]

There is a strong invitation here to attribute current validity
to this prescription for good theater acoustics. Certainly the
faults of excessive reverberation and echo are no less to be
avoided now than when Vitruvius wrote. It must be remem-

bered, however, that the type of theater for which Vitruvius is discussing the proper location is what would now be called an open-air amphitheater. Vitruvius would have been entitled to even more concern over the case-endings had he envisaged the compounding of acoustical difficulties introduced by adding walls and a roof—additions which became necessary, of course, when theaters were built in less temperate climates than that of his native Italy. Moreover, Vitruvius' reference to the interference between directly transmitted and reflected sound can hardly be taken to indicate any understanding of the modern concept of wave interference. Neither can his reference to sound which "dissolves" and dies away with "indistinct meaning" be counted as a very accurate description of reverberant sound decay. Nevertheless, Vitruvius must be given high rank as an acoustical prophet for his identification, if not for his clear understanding, of a few of the difficulties whose avoidance is still a prime objective for practitioners of architectural acoustics.

It is interesting, in view of the modern military exploitation of acoustical location techniques, to find that the earliest record of the practical application of acoustical principles turns up in a similar connection. (Unless, that is, the trumpeting down of the Walls of Jericho[53] can be counted as the first example of applied acoustics.) The military problem in question was the location of tunnels being driven under the walls of a beleaguered city. At the close of the sixth century B.C., the Persians laid siege to the small Libyan town of Barce (= Barca, now called El-Merjeh) and sought to breach the walls by tunneling under them. In describing this campaign, Herodotus (ca. 484–425 B.C.) tells how "a smith discovered them [the tunnels] by means of a shield coated with bronze, and this is how he found them: carrying the shield round the inner side of the walls he smote it against the ground of the city; all other places where he smote it returned but a dull sound, but where the mines [tunnels] were the bronze of the shield rang clear. Here the Barcaeans made a countermine and slew those Persians who were digging the earth."[54]

The last book of Vitruvius' treatise, devoted largely to the construction of various engines of war, describes another method of tunnel location which displayed somewhat more acoustical sophistication. Philip V of Macedon conducted a campaign in 214 B.C. against the town of Apollonia (now Pollina), in Illyria, just across the Adriatic from the heel of Italy's boot. Spies brought to the defenders the disconcerting news that the Macedonians "designed by digging tunnels to penetrate unsuspected within the walls." Vitruvius then ascribes to the architect Trypho of Alexandria the conception of a plan of defense comprising countertunnels driven beneath the walls, in which "Everywhere he hung bronze vessels. Hence in [the] one excavation which was over against the tunnel of the enemy, the hanging vases began to vibrate in response to the blows of the iron tools. Hereby it was perceived in what quarter their adversaries purposed to make an entrance with their tunnel."

Surely this was a splendid exploitation of the phenomenon of resonance under shock excitation, made centuries before the physical phenomena involved were to be understood. But if the method of detection was advanced, the retaliatory measures taken were certainly primitive enough to satisfy the most vindictive. "On learning the direction," Trypho proceeded to fill "bronze vessels with boiling water and pitch overhead where the enemy were, along with human dung and sand roasted to a fiery heat. Then in the night he pierced many openings, and suddenly flooding them, killed all the enemy who were at work there."[55a] One can only conclude that war is grim in any age!

In his discussion of the proper training of architects, Vitruvius lays it down that "a man must know music that he may have acquired the *acoustic* and mathematical relations and be able to carry out rightly the adjustments of *balistae*, *catapultae*, and *scorpiones*." This requirement is explained further in connection with his description of the construction

and stringing of these primitive "guns," where it is specified
that the ropes are to be "coiled round the windlass, so that
when the ropes are stretched by the levers, and struck by the
hand, they may respond with the same sound on either side.
. . . So catapults are tuned to the right note by the sense of
hearing for musical tones."[55b] This, presumably, is the historical antecedent for that delicate acoustical perception by
means of which a skilled mechanic makes the final adjustments—which he too calls "tuning up"—on the engine of a
racing car, boat, or airplane.

The Role of Observation and Inference

The urge to partition history into eras that can be neatly
tagged for identification has reserved the designation "Age of
Experiment" for attachment to a later period (sixteenth
century on). However, the contrary implication that no deliberate experiments were performed in the period of antiquity
can hardly be supported. To be sure, when pure reason and
experience appeared to be at odds, Plato and Aristotle showed
no hesitation in choosing to accept the conclusions based on
reason as the true reality. It is also true that the political
climate of their times made dictatorship a familiar burden,
and the authoritarian influence of Aristotelian doctrine became a dominating factor in the development of natural
science for many centuries. Nevertheless, to attribute to the
age of authority a universal disdain for experiment is to do a
gross injustice to the lively sense of curiosity exhibited by
many of the primitive natural philosophers. Even though they
did not perform very many experiments of the kind later
designated as planned or controlled experiments, they certainly did bring a keen sense of observation to bear on the
physical world about them.

There are abundant examples of the techniques of observation and inference as practiced in the pre-Christian era. The
conclusions drawn by the early Pythagoreans about vibrating
strings were certainly based on observation; and even if the

monochord was not designed deliberately for the purpose of experimentation, it was certainly used for this purpose. It has already been pointed out that the interplay of observation and inference was an essential feature of the "appeal to . . . hearing and intellect" advocated by Aristoxenus and his followers. And although Aristotle could choose reason over experience in extremis, he was certainly motivated in many of his excursions into natural science by a desire to apply reason to experience. His conclusion that light bodies fall more slowly than heavy bodies is often cited as evidence of a lack of experimentation—for did not Galileo find a contrary result when he did try the experiment? It is perhaps more equitable to credit Aristotle with having reached this conclusion through a deliberate attempt to develop a mathematical theory of dynamics. He *did* reach the conclusion that all bodies would fall alike in a vacuum; but since, on his assumptions, the common velocity of fall in the unresisting void would be infinite, he chose to use this unpalatable consequence as proof that a vacuum cannot exist. This tangled web of the true and the false may well be one of the best, but it is certainly not the last, example of mixed results flowing with faultless logic from a mixture of true and false assumptions.

It is perhaps significant that the most outstanding examples of observation and inference at work in antiquity are to be found in fields other than sound. That there are good and sufficient reasons for this was suggested in the Introduction, and these reasons will deserve and receive repeated consideration as this discussion proceeds. In the present context it is enough to point out that even the elementary aspects of mechanics, heat, and light could only be dealt with on a phenomenological basis that was inherently superficial. The anatomical miracles of speech and hearing, no less than those of vision, had to be dealt with in terms of the projection and perception of something. Thus "voice" had to be given a corporeal entity in order that it might "travel" and "echo," and behave generally in an understandable way. There was certainly ample justification for the ancients to be preoccupied with the basic elements of physical science, and little room in their scheme of things for the appearance even of precursors of a point of view that would combine wave concepts adequate to deal with optical phenomena with the mechanics of a compressible material medium.

As an illustration of the foregoing, both experience and experiment sustained a lively interest in the branch of optics dealing with reflection and image formation in mirrors (i.e. *catoptrics*). This subject was dealt with extensively by Heron of Alexandria (fl. second century B.C.), an applied scientist who is best remembered in this age of jet propulsion for his invention of the aeolipile,[56] the first steam reaction "engine." Heron held, in common with Plato and Euclid, the ancient misconception that light proceeds from the eye via the mirror to the seen object, but this inversion did not vitiate the proof he gave that the angles of incidence and reflection are equal. Although this principle had been established earlier, Heron's method of proof[57] is of considerable interest, since he clearly anticipated Fermat's principle by basing his argument on the assumption that the total path traversed by the light is a minimum, a principle that was to be recognized much later as applying with equal validity to sound and to light.

The great Nicaean astronomer Hipparchus (190–125 B.C.), a predecessor of Ptolemy, was one of the first observers of the heavens who made accurate, systematic measurements. He was also a first-rate mathematician, and is credited with laying the foundations of trigonometry and with inaugurating the use of latitude and longitude as a method of designating fixed positions on the earth. His greatest astronomical achievement was to discover, and measure with creditable accuracy, the precession of the equinoxes; but he also made a catalogue of 850 or more fixed stars classified in six categories of magnitude, and determined the periods of all the planets then known and the relative sizes and distances of the sun and moon.

The circumference of the earth had been determined[58] earlier by Posidonius (ca. 135–50 B.C.) and by Eratosthenes (ca. 275–194 B.C.), the latter not only having done it first but with more accuracy. Unfortunately, when Ptolemy needed this datum for his map of the world, he either repeated the measurement wrongly or else adopted without criticism the too-small estimate made by Posidonius. As a result, Ptolemy's map systematically underrated east–west distances by a gross error that increased toward the outer edges of his map. And since it was to be a long time before another geographer of Ptolemy's distinction appeared, it was possible for these erroneous data to turn up with modifications fourteen centuries later in the hands of the Florentine physician and naturalist, Paolo dal Pozzo Toscanelli, a teacher of the young Leonardo da Vinci, and to be further propagated by Toscanelli in letters to Canon Martins and to Christopher Columbus that were influential in persuading the latter to undertake his westward voyage of discovery.[59]

Heron did much to put the engineering and land surveying of his day on a scientific basis, and his dominant interest in the applications of science led him to the invention of many ingenious machines and toys. However, the ancient most entitled to be honored as the patron saint of mechanical engineering science is probably Archimedes (287–212 B.C.)— though he would probably execute an Archimedian spiral in his grave at the thought of such recognition, since he shared Plato's conviction that it was both improper and unseemly for a philosopher to seek to apply the results of science to practical use. The modern scientist called upon to divert his talents to military ends can take some comfort in knowing that, in spite of his idealism, Archimedes was similarly diverted; for he was a prolific inventor, and his advice was often sought by his government concerning difficulties that could be overcome by material means. He devised the catapults that helped keep the Romans at bay during their siege of Syracuse, and some interesting but questionable accounts give him credit for constructing a large burning glass, comprising a composite mirror, with which ships of Marcellus' Roman fleet were alleged to have been set on fire at a considerable distance. It is more certain that he invented the compound pulley and the Archimedian screw, which still serves in many instances as a convenient method of raising water. Archimedes' celebrated discovery of the principle of displaced volume (*Eureka!*) and his brilliant combination of this simple principle with that of weighing to measure density, and thus to detect counterfeiting in King Hieron's crown,[60] constitute classic examples of carrying nature's question to nature.

Assessment and Retrospect

The foregoing typical examples of discovery and invention make it hard to deny the existence in ancient times of a vigorous interest in science and the fruits of science. And yet, with the wisdom of hindsight, it is easy to find cause to lament the sins of omission of the primitives, their missed opportunities to turn perception into conception, or even to know when they had made a valid observation. The commentators and historians who repeated the legend of Pythagoras and the hammers could so easily have contributed to the store of knowledge in their own time had they only repeated the simple experiment of listening to some hammers and then weighing them. In fact, most of the tools and instrumentalities of science that were available to Galileo and Mersenne could also have been employed by Archimedes and Aristoxenus; yet observation remained predominantly casual instead of directed, and experimentation was more often opportunistic than deliberate.

Perhaps the most notable feature of this period of antiquity was, to put it in the vernacular, the build-up for a big letdown. The Golden Age was glittering indeed but only its marbles and its ethics survived. In the area of natural science there was almost none of the stabilizing continuity that flows from the compounding of new knowledge on a firm foundation of the old. Each succeeding school of philosophy displaced the one before, either by evading its issues or by overwhelming it with prevailing if not superior disputation, and progress in understanding was as often circular as forward. Only in mathematics was it possible, as Roger Bacon put it nearly a millennium later, "to arrive at the full truth without error, and at a certainty of all points involved without doubt; since in this subject demonstration by means of a proper and necessary cause can be given."[61]

With regard to matters that concern physics, which is to say with regard to matter itself, there was perversely absent any acceptance from preceding generations of a conceded body of established facts. Aristoxenus could declare boldly, "We shall advance to our conclusions by strict demonstration,"[62] but there was no answering surge of vigorous curiosity, and one must conclude that the time was not yet ripe for sustaining a true spirit of scientific inquiry. Without it, the onset of stagnation and its transition to decadence were inevitable. Both the will and the right to inquire, the mark and herald of the scholar, withered with the decline and all but perished with the fall of Rome; and the Western world turned to occupy itself for more than a thousand years with politics, strife, conquest, and the other arts.

2 ORIGINS IN EXPERIMENT: GREGORY I TO NEWTON

The second period of this chronicle of acoustics, like the first, extends over nearly ten centuries. It might be characterized alliteratively as the period from Boethius to Boyle or, to widen the limits slightly, from Nichomachus to Newton. The geographical trajectory of influence, however, did not follow a simple geodesic from Rome to London. On the contrary, while the dogs of war fought over the bones of Europe during the so-called Dark Ages, the lamp of science burned low in the West and might have gone out had the flame not been kept burning, like an Olympic torch, in the Middle East. The Golden Age of Greece and the glory that was Rome had their counterpart in the Golden Age of Islam that reached its climax during the tenth and eleventh centuries. These "middle" ages were only "dark" in the Christian West.

The pendulum swings wider, perhaps, during this period than for any like interval in man's history. Geographically, the trail leads from Athens to Rome and back to Medina, from Baghdad to India and return, and through North Africa to Spain and thence to England and western Europe. The Byzantine school, the Hindus, and especially the Arabs, served posterity in a vital way during these times, by acting as conservators, nourishing the Greek scientific tradition and preserving it relatively intact. As in the parable of the man who buried his talents, however, the stewardship of this heritage of natural science reverted to the West after the intellectual sterility of the Dark Ages had released its hold on western Europe.

Not many of the documents embodying the Greco-Roman tradition survived the rigors of conquest and migration in their original form, but the Greek scientific tradition itself did, thanks to wholesale translation of the Greek manuscripts into Arabic during the eighth and ninth centuries and their retranslation into Latin during the twelfth and thirteenth. The practice of commissioning the translation into Arabic of as many of the Greek and other "foreign" scientific works as possible was instituted by the second 'Abbāsid caliph Al-Mansūr (ca. 710–775), and was continued by Harun Al-Rashīd (763–809) and by the greatest of the 'Abbāsid monarchs, the caliph 'Abdallāh al-Ma'mūn (786–833).[1] The latter even went so far as to send a military mission (ca. 813) to the Byzantine emperor Leon of Constantinople to demand both ancient documents dealing with art and science and Greek scholars to assist with their translation.[2] Under such conditions of active, not to say demanding, patronage, it is not surprising that some of this work of translation was incompetently done; but a good bit of it survived long after the originals had been lost, and many important classical writings are now known only through these copies.

It would be a mistake, however, to overemphasize the passive role of conservation in medieval science. The dissemination of Greek scientific ideas throughout Islam was never uniform or complete, and a substantial number of Muslim scientists are to be credited with making independent discoveries quite comparable in significance to those made by Greek scientists. The fact that relatively few of their discoveries actually contributed understanding very much beyond that achieved by the Greeks, especially in the field of physical science, does not at all detract from the praise owed these original thinkers. Moreover, and not surprisingly, the advances most worthy of note were those associated with the rise of interest in, and use of, deliberate experimentation—the distinguishing feature of modern science that had been so conspicuously missing in much of Greek science.

Acoustics versus Scholasticism

The rise and decline of medieval scholasticism carried the pure rationalism of Greek natural philosophy through a full cycle of change before free scientific inquiry was reinstituted in the West in the culmination of the Renaissance. Three of the five religions were already flourishing during Greece's Golden Age. Amos, the earliest known of the Hebrew monotheistic prophets, had been active during the period 765-750 B.C.; and both Confucius (551-479 B.C.), and Gautama Buddha (ca. 560-ca. [483?] 477 B.C.) had been contemporaries of Pythagoras. None of the followers of these faiths, however, played an important role in the development of Greek science, and there were no artificial religious or philosophic barriers for Aristotle or Archimedes to surmount except those of their own creation. But, alas, a profound and irreversible change in this situation set in following the birth of Christianity[3] in the West and the rise of the Mohammedan faith (= Islam: the prophet Muhammad, ca. 570-632) in the Middle East.

It is easy to see why there was bound to arise, sooner or later, a fundamental clash between religious faith and the objectivity of scientific rationalism. One is static and invariable, the other is dynamic and committed to change; yet each is responsive to deeply felt human needs. The responsibility for fusing these inhomogeneous convictions fell then (as it still does) on the shoulders of teachers and scholars, and their attempts to achieve a synthesis of sacred and profane sciences were a dominant feature of the scholasticism that flourished almost universally during medieval times. Moreover, since education, even at the university level, did not escape the domination of the church until late in the Renaissance, it is not surprising that the scholastics of every faith seemed able to achieve, each by his own standards, an equally satisfactory fusion between theology and technology.

The broad implications of what Sarton called "the cause and cure of scholasticism"[4] may seem to be strange grist for the acoustical mill. There were at least two reasons, however, why the branch of acoustics dealing with music was able to make a unique contribution toward the ultimate conquest of scholasticism. The first was that music was assured a firm continuing hold on its place in the scholastic sun by virtue of its role as a part of the classical quadrivium. Educators, philosophers, encyclopedists, and commentators alike had perforce to deal with music and with the evolution of musical science. The second basis for the close relation between music and scholasticism stemmed from the fact that music is, sui generis, an epitome of experimental science. Objective in execution and humanistic in appreciation, its three aspects of *composition, performance,* and *appreciation* exemplified—and held up continuously for conscious or unconscious regard—the scientific credo of *hypothesis, experiment,* and *conclusion.*

If the attention devoted to music had been wholly concentrated on its esthetic aspects, it would be more difficult to support the importance here attributed to it. Happily for the purposes of this study, however, almost every medieval writer who considered the theory of music felt obliged to devote at least one section of his treatise to the production of sound and to the factors that influence pitch. These discussions amply support the contention that understanding of the physics of sound production did not advance significantly beyond the stage achieved by the Greeks until very near the end of the medieval period. Indeed, it could hardly have been otherwise, since the Greeks had already achieved an understanding of the physical nature of sound that represented almost as much sophistication as could be sustained without explicit formulation of the laws of dynamics and of the physical nature of the sound medium. It was clearly demonstrated, however, that the Greek tradition in physical acoustics did become in due course fully assimilated by Muslim scientists; that many of the false ideas expressed along with the true by the Greeks were weeded out; and that a clarified, if not augmented, version of the Greek acoustical tradition was conserved in Islam for retransmission—"on the wings of song," as it were—to the West.

The Musical Thread of Continuity

The first name to reckon with in connection with the development of musical theory is that of Gregory the Great (540–604), who became Pope Gregory I in A.D. 590. He made many important contributions to church music but he is best remembered for establishing the form of melodic line now identified as the Gregorian chant.[5] Prior to Gregory's time, the notation used in writing music did not include any symbols for indicating the relative duration of successive notes, this being left to the free choice of the performer. Gregory took the first halting step toward the concept of "mensural" music by assigning equal duration to each note of his simple melodies. The term *mensural music* was only introduced, however, after this concept had been broadened by the introduction of melodic components having different and specifically designated time duration, or "length." Of course, as musical notation evolved toward its present form, all music became mensural; but the written language of music was still in a primitive state in the eighth century. The Arabian lexicographer Khalīl Ibn Ahmad (b. 717, fl. at Basra, d. 791) seems to have been the first to write on this subject, but unfortunately his *Kitāb al Īqā* (Book of Rhythm) is no longer extant and his contribution to mensural music can only be judged by the acknowledgments of his influence by later theorists.

The Arabian musical tradition[6] was carried on during the first half of the ninth century by Al-Kindī (Alkindus) (b. ca. 800, fl. in Baghdad 813–842, d. ca. 874), one of the great

philosophers of medieval times. Al-Kindī was well versed in
Greek science and philosophy, having himself translated many
Greek writings into Arabic. In turn, some of his own writings
were eventually made available in Latin by the twelfth-
century translators, with the result that his influence continued to be significant during the later medieval period when
interest in natural philosophy began to revive in the West.

Other translations from the Greek, including some treatises
on music, were either made or commissioned in Al-Kindī's
time by the three brothers, Banū Mūsā (fl. mid-ninth century), each of whom was both a scientist and a patron of
science. In the latter half of the ninth century this process of
transplantation continued and more tracts on music[7a] appeared from the pens of various Arabs not distinguished
otherwise for their contributions to either sound or music.
Among these authors and translators can be mentioned the
disciple of Al-Kindī, Al-Sarakhsī (fl. ca. 890, d. 900), and
three others who flourished at Baghdad, the mathematician
and astronomer Thābit ibn Qurra ([827?] 836–901), the
Christian mathematician Qustā ibn Lūcā (d. ca. 912), and the
great clinical physician Al-Rāzī (Rhazes) (ca. 850–924).

During this same period the Christian West could offer only
the two Benedictine monks Regino of Prüm (d. 915), a German historian best known for his chronicle of the first nine
centuries of Christendom, and Hucbald (840?–930). The latter proposed an alphabetical notation[7b] and was one of the
first to deal with polyphonic music, but each is credited with
a musical treatise bearing the title *De Armonica institutione*.
Except for Hucbald and Al-Kindī, none of these early medieval authors can be credited with any significant original
contribution, and it is reasonable to infer that activity in this
field during the ninth century was largely confined to making
available in Arabic the Greek acoustical tradition.[8]

Very much more originality was displayed by Al-Fārābī
(Alpharabius) (ca. 870–950), a Muslim philosopher and mu-

sician who, like Al-Kindī, was conversant with most of the
scientific thought of his day. A minor treatise on *The Rise of
the Sciences* is attributed to him (though not wholly without
question)[9] and is of particular interest, not only because it
mirrored the science of his time, but also because it revealed
the growing inclination and readiness to bend nature deliberately to useful purpose. In speaking of the past and future
scope of natural science, Al-Fārābī said, "We need a science
which deals inclusively with changes in nature, showing how
[such changes] are brought about and their causes, and enabling us to repel their harmful action when we wish, or to
augment them,—a science of action and passion."[10] His own
willingness to augment change was to be shown in his contributions to music.

With regard to the physics of sound production, Al-Fārābī
displayed an understanding that was largely Aristotelian—and
not very late Aristotle at that, since he still held some misconceptions that had been cleared up within Aristotle's time.
For example, Al-Fārābī declared, in a style not very different
from that of Strato's *De Audibilibus,* that "It is the air thrust
away by the impact of two bodies which transmits the sound.
It moves the layer of air immediately adjacent with a motion
similar to its own; the latter communicates this motion to the
next layer, and so forth. The sound being in this way transmitted from one layer of air to another, reaches the air contained within the auditory tract, then the organ which is the
seat of the auditory faculty." He drew a distinction between
note and sound in general by specifying that "The note is a
sound that persists for an appreciable interval within the
body in which it is born." His further explanation of this idea
completes a passage that suggests an awareness, if not an understanding, of resonance and of wave propagation: "Bodies
that are capable of vibrating produce notes. We mean by that,
those [bodies] which, having received motion in different
directions, conserve this motion for a time. The initial motion

extends progressively to all parts of the body, all its molecules, even after the initial cause has disappeared. Strings are examples of this kind of body."[11a]

On the other hand, Al-Fārābī clung tenaciously to the misguided pre-Aristotelian notion that "the highness or lowness [of the pitch] of sound generally depends on the degree of compression given to the molecules of the layer of air which rebounds under impact," and to the equally mixed-up idea that "the harder and smoother the body which is hit, the higher the sound produced."[11b] With regard to the influence of length and tension on the pitch of "notes" produced by vibrating strings,[11c] Al-Fārābī was soundly Pythagorean; but his rationalizations about the pitch of flute tones were relatively wide of the mark, owing in part to his confused understanding of the independence of intensity and pitch, and in part to a similar confusion between the velocity of sound and the velocity of the air impelled by the player's breath.[11d]

In the field of music, Al-Fārābī showed up to much better advantage. Both Al-Kindī and Al-Fārābī were familiar with the concept of mensural music and the latter dealt with it extensively in his grand treatise, the *Kitāb al-Mūsīqī*.[11e] In addition, each had other contributions to make to musical theory. Al-Kindī employed a novel musical notation that provided in a rudimentary way for indicating both tone duration and pitch, while Al-Fārābī recognized and made use of the consonances of the major third (5:4) and minor third (6:5) in two-part music.[11f] Notice, though, that while Al-Kindī's notation could specify the pitch by word or symbol, there was no universal physical standard yet for its determination. Indeed, it was fortunate for musical theory that only the ratios of frequencies had to be known in order to determine pitch intervals. These it had been possible to identify ever since Pythagoras—for example, in terms of the ratio of the lengths of two segments of a vibrating string; but finding the absolute numerical value of the frequency associated with any particular note proved to be a more difficult problem that was not solved until the seventeenth century. Al-Fārābī's "recognition" of the major and minor thirds as consonances was of greater significance than might appear at first. In Pythagorean intonation, these intervals were dissonant—a shortcoming of the Pythagorean scale that Aristoxenus and Ptolemy had each sought to remedy. It follows that Al-Fārābī's acceptance and use of these intervals constitute circumstantial evidence that at least some modification of the Pythagorean scale had already attained a degree of currency by the tenth century. It is also suggestive to consider the relationship between Al-Fārābī's use of these intervals in two-part performance and the growth of the modern harmonic theory of musical composition.[12] Even before the time of Pythagoras, Terpander (fl. ca. 710 to 670 B.C.) had made use of "doubling with the octave" in musical performance. This practice came to be called *magadizing*, which amounted to simply singing or playing a melody with two "voices" always separated in pitch by exactly an octave. The contribution made by Al-Fārābī, and perhaps earlier by Al-Kindī, was to introduce doubling also with the consonant intervals of the third, the fourth, or the fifth. This was called *organum*, and its practice was referred to as *organizing*.

The importance of organum as a stage in the evolution of music can hardly be overemphasized. Prior to its introduction in the ninth and tenth centuries, a theory of the concord of *simultaneous* sounds, or musical *harmony* as this term is now understood, was virtually nonexistent. Except for the Grecian magadizing, with its fixation on the consonance of the octave, all studies and theories of musical concord had dealt with *successive* sounds. Obviously the scope for self-expression in musical composition was enormously enlarged by the possibility of introducing harmonic variety, as afforded first by the diaphonic practice of organizing.

It seems equally obvious (now!) that the next stage of

musical evolution was almost sure to be one that would free the two voices from the requirement of following exactly the same *rhythmic* pattern. This innovation was probably of Arabian origin and took initially the form of a "gloss" or ornamentation of the melody, known as *descant*—a form of two-part performance in which one voice rendered the melody in plain form while the other was permitted to embellish it with interpolated grace notes and other variants. No firm dates can be given for the introduction of any of these innovations, since it is likely that they had their primary origin in the anonymity of casual musical performance. However, if they did not originate in Muslim music, it can at least be said that they were eventually transmitted to the West through this channel.

It has been said that the basis of any musical theory is the definition of the modes and genres—terms that deal with the sequences of *pitch* intervals comprising the various major and minor scales. In fact, prior to the lost *Book of Rhythm* by Khalīl Ibn Ahmad, this was almost the only problem of musical theory, but it was quite adequate to sustain a lively spirit of contention among the Greek scientists who devoted themselves to it. Ptolemy's proliferation of the modes had merely added fuel to these fires of contention, which were still burning after several centuries.

An articulate concept of identifiable *rhythmic modes* was much slower in developing, in spite of the fact that rhythm itself has always been an essential feature of musical performance. Songs were sung to be understood and felt; and since they were usually poetry, it was easy to persuade the performer to adopt the rhythm of the poetic meter. It followed, however, that as mensural music developed at the hands of theorists who followed Khalīl and Al-Kindī, it fell heir in due course to all the rigors and complexities of versification and prosody. In spite of these growing pains, exciting new fields for musical invention were opened by the evolution of a mensural notation adequate to deal with the rhythmic dimension, especially in connection with instrumental music. With regard to song, these developments had the further effect of shifting some of the burdens of responsibility from the lyricist to the composer, and thus of giving birth to the still flourishing interplay of dominance between words and music.

The ethos and emotional content of music, and the relation of music to morals and politics, have been subjects of concern from ancient times down to the present. Plato and Aristotle discussed the various modes from this point of view, and the early Christian church was not indifferent to the unsuitability of some modes for church music.[13] In Muslim hands the voice of song began to acquire new power—and to become correspondingly more suspect. It was for such a reason, presumably, that the line of orthodox caliphs was led during the early days of Islam to promulgate the doctrine Music and All Instruments Are Forbidden to the Faithful, an interdiction expressed in words attributed to Mahomet: "To listen to music is an offence against the law; to make music is an offence against religion. To take pleasure in it is an offence against the faith, and renders one guilty of infidelity. Music is not permitted either in private or in public, nor in any circumstance of life (except for the musical call to prayer), not even for weddings."[14]

Surely there would have been very few of the truly faithful in Islam if these precepts had been rigorously sustained, but a wise counsel of moderation seems to have prevailed even before orthodoxy was reinterpreted. An engaging discussion of the niceties of this question was given by the Arab theologian Al-Ghazzālī (Algazel) (b. 1058, fl. at Nishābūr and Baghdad, d. 1111) in his *Iḥya 'ulūm al-dīn*. He concluded finally that

listening to Music and Singing is sometimes absolutely forbidden and sometimes permissible and sometimes disliked and sometimes to be loved. It is forbidden to the most of mankind, consisting of youths and those whom the lust of this world controls so that Music and singing arouse in ˙them only . . . blameworthy qualities. . . . It is

allowed with reference to him who has no delight in it except taking
pleasure in beautiful sounds. . . . And it is loved with reference to
him . . . in whom Music and Singing arouse only praiseworthy qualities.[15a]

This measured judgment is hardly as exciting as Al-Ghazzālī's
lyrical testimony concerning the effects of music on man:

> Lo! Hearts and inmost thoughts are treasuries of secrets and mines of
> jewels. . . . There is no way to the extracting of their hidden things
> save by . . . listening to music and singing, and there is no entrance to
> the heart save by the antechamber of the ears. So musical tones,
> measured and pleasing, bring forth what is in it and make evident its
> beauties and defects[15b] . . . and as for Mālik (may God have mercy
> on him!), he has forbidden singing.[15c]

Although such stylized invocations were as common in
Arabic writing as *amen*s at a revival meeting, one can infer
that Al-Ghazzālī felt that Mālik stood in special need of such
mercy! Al-Ghazzālī made it clear, incidentally, that he had
some knowledge of mensural music when he explained that

> pleasant sound . . . is divided into measured and not measured; and
> the measured is divided into what has a meaning to be understood,
> such as poems; and what has not, such as the sounds produced by
> lifeless substances and by other animals than men.[15d] . . . Some
> sounds make to rejoice and some to grieve, some put to sleep and
> some make to laugh, some excite and some bring from the members
> movements according to the measure, with the hand and the foot
> and the head. And we need not suppose that [this comes about]
> through understanding what is meant by the poetry, for it is pos-
> sible in the case of stringed instruments

to which Al-Ghazzālī added, in ominous benediction, "He
whom the spring does not move with its blossoms, nor the
'ūd [lute] with its strings, is corrupt of nature; for him there
is no cure."[15e]

The importance and dignity accorded to the problem of the
legality of song reflects properly the deep embedment of
music and song in Muslim life. Almost every phase of activity
was touched by it. Slaves who could sing brought a higher

price on the market; with song, camel drivers alerted and urged their charges on and sheepherders turned their flocks. Fishermen dug holes in the marshes of Iraq around which they gathered to utter stirring sounds intended to attract fish. Even physicians were ready to concede, according to one tenth-century commentator, "that the beautiful voice moves in the body and flows in the veins. In consequence, the blood becomes pure through it, and the heart is at rest through it, and the soul is quickened through it, and the limbs are agitated, and the movements are brisk."[16]

In spite of these instances of "applied music," however, it was songs of the heart that appealed most deeply to these simple people of ready emotion. Two splendid literary monuments of this golden age of song were the *Kitāb al-Aghāni* (Book of Songs)[17] written by Abu'l-Faraj al-Isfahānī (b. 897, fl. at Baghdad, d. 967), and the *Murūj al-dhahab* (Meadows of Gold)[18] written by Al-Mas'ūdī (ca. 912–ca. 957). Perceval has made available a somewhat less pretentious series of biographical anecdotes,[19] drawn from the same period, which combine much of the charm of the *Canterbury Tales* and the flavor of the *Arabian Nights*.

The Sound of Science in the Middle Ages

Late in the tenth century (ca. 893) there was established at Basra a secret society called the Ikhwān al-Safā (Brethren of Purity). This group, which more nearly resembled one of the eclectic schools of Greek philosophy than it did a modern scientific society, collaborated in the compilation of more than fifty tracts dealing with every phase of philosophy and natural science known in their time. On the whole, their doctrine resembled the Pythagorean and Platonic as much as it did the Aristotelian. Thus the "harmony of the celestial spheres" made another appearance, and there was a "note" and a "melody" befitting every temperament and nature.

The acoustics portion of the Ikhwān al-Safā tract on mathematics[20] contains rather more that is naïve and fanciful than

Al-Fārābī's treatise did, but it also reveals a few conceptions of distinctive originality. Sounds are, as usual, "generated in the air by the collision of objects."[20a] The Ikhwān avoided Al-Fārābī's errors in discussing pitch and intensity, but they would still have one believe that "the tones of smooth objects are smooth because the interfaces common to them and to the air are smooth."[20b] The notion of spherical wave propagation—first introduced by Chrysippus (see part 1, above) and later taken up by many Latin commentators—was reintroduced with a new three-dimensional figure of speech: "The air between them [the colliding bodies] is compressed, the particles composing it collide and disperse in a wave-like motion in all directions; from this motion originates a spherical configuration which spreads, in the way a bottle widens as the glass blower inflates it with air. As this configuration widens, its motion becomes weaker and oscillates until the motion subsides and ceases."[20c]

The sophistication revealed by this statement is somewhat deflated by the remarks that follow it: "If a human being or some other animated being provided with ears is present and near this location the oscillating air enters its ears and penetrates to the two ear cavities in the rear part of the brain. [!] The air in that region oscillates also and the auditory sense perceives this motion and this change."[20d]

The Ikhwān appear to have been among the first to distinguish, albeit clumsily, the attributes of sound designated by intensity, pitch, and quality. "Tones are strong or weak; rapid or slow; fine or thick; heavy or light,"[20e] they said. The terms "fine or thick" seem to have carried the high- or low-pitch information, and were perhaps suggested by the nature of the corresponding strings on the lute. "Rapid or slow" was explained in terms of the "succession of impacts," so this property may also have applied specifically to the sounds of plucked strings. On the whole, however, these were competent observations. In still another prescient observation that bears on the concept of the total power of a sound source, the Ikhwān told how "Hard, hollow objects like pots, beakers and jars sound for a long time after they are struck because the tone repeats itself in their cavities and collides again and again until it comes to rest. Those [hollow vessels] which are wider give a stronger tone because they push more air inward and outward."[20f]

Muslim science reached the summit of its achievement in the late tenth and early eleventh centuries at the hands of three natural philosophers of outstanding quality. These were the Egyptian Ibn Al-Haitham (Alhazen) (ca. 965–1039), and the two Persians Al-Bīrūnī (973–1048) and Ibn Sīnā (Avicenna) (980–1037). No contributions to sound in general or to music have survived in the writings of the first two, but their influence on other branches of science entitles them to respectful notice. The physicist Ibn Al-Haitham flourished at Cairo under Fāṭimid patronage and devoted himself especially to research in geometrical, physical, and physiological optics. He carried these subjects far beyond the point reached by the Greeks, and his great work on optics (*Kitāb al-manāzir*), which was hardly overtaken until the sixteenth century, was said to have influenced Bacon and Kepler and to have provided the basis for Witelo's famous thirteenth-century treatise on perspective. Ibn Al-Haitham was also credited with writing a commentary on Euclid's *Sectio Canonis*, as well as a discourse on *Harmonics*, but unfortunately neither of these manuscripts is extant.

Al-Bīrūnī displayed somewhat broader interests than Ibn Al-Haitham, and his extended travels in India allowed him to forge an important link between Muslim and Hindu science. His writings were distinguished by a critical spirit and intellectual courage worthy of Galileo and unparalleled in his time. Thus he not only believed, but dared to press the claim, that the stereotyped phrase "Allah is omniscient" did not constitute a justification for the complacent acceptance of individual ignorance. The Muslim credo contained many similar rationalizations that drew Al-Bīrūnī's fire, one of which

can be illustrated anecdotally by the apocryphal legend of
the burning of the remnants of the great library at Alexandria
following the second conquest of that city by the Caliph
Omar in ca. 646. When this incipient vandalism was protested,
the caliph is said to have replied, "If these books contain
what is agreeable with the Book of God, the Book of God
is sufficient without them; if they contain what is contrary
to the Book of God, there is no need for them; so proceed
with their destruction."[21] This was the kind of sophistry that
Al-Bīrūnī dared to challenge. His writings do not have very
much to say about sound, but he is said to have revived, on
the basis of his own observations, the awareness that the
speed of light is incomparably greater than that of sound.[22]

The greatest of the Muslim philosophers was Ibn Sinā, a
scholar of almost universal interests and still called in Islam
the "prince of all learning." One of his most important
writings was a philosophical encyclopedia (Kitab al-Shifā)
which included, among many other things, a section on sound
and music. A less well-known manuscript entitled *Kitab al-
Nafat* throws further light on his understanding of and contribution to musical theory.

Ibn Sinā divided all theoretical knowledge into the categories of physics, mathematics, and metaphysics; and all
practical knowledge into ethics, economics, and politics. He
wrote on all these topics in the Aristotelian tradition, modified somewhat by Neoplatonism and influenced by Muslim
theology. Of course, to identify such a mixture of influences
is tantamount to crediting Ibn Sinā with the establishment of
his own distinctive philosophical viewpoint. He made a study,
profound for his time, of such physical questions as those
concerning motion, force, the vacuum, light, and heat. He
stood almost alone in denying the possibility of transmuta-
tion, holding that the differences between metals were not
superficial ones; and he made the further observation that if
the perception of light is due to the emission of some sort of
particles by luminous bodies, the speed of light must neces-

sarily be finite. Ibn Sīnā also wrote a monumental medical encyclopedia, the Qānūm, which was similar to Galen's classical second-century treatise but surpassed it in some respects, and largely superseded it. The authority of the Qānūm was almost supreme for the following six centuries, and it must still be counted as an influential treatise throughout Islām.

In his writings on sound and music, Ibn Sīnā's explicit discussions of *magadizing* and *organizing* carried this art considerably beyond its beginnings in Al-Fārābī's hands. Ibn Sīnā also considered in some detail the series of consonances represented by pairs of tones in the superparticular ratio $(n+1):n$. He observed that when $n = 33$, the interval becomes so small that the two tones sound nearly alike in pitch, and that when n is as great as 45, the ear is unable to distinguish the two tones.[23] These observations constitute a remarkable and early example of deliberate experimentation in acoustical science, and they add to the record the earliest measurement of the differential threshold for pitch change. It redounds to Ibn Sīnā's credit, moreover, that his experimental result is in very good agreement with contemporary determinations of this threshold.

The Rustle of Revival in the West

The first stirrings of scientific revival in the West began to appear during the eleventh century. On the acoustical front, the most significant contribution was probably made by Guido of Arezzo (Aretinus) (ca. 990–1050), who was responsible for major reforms in the methods of teaching and writing music. Although Guido is not to be credited with the invention of the musical staff, he enhanced its serviceability by making use of both the lines and the spaces between the lines to indicate the position of notes in the scale. It is a warmly debated question whether the Arabs or Guido first proposed to designate the notes of the scale by spoken syllables.[24] Popular but unverified legend has it that Guido was the first to adopt the sequence of phonemes comprising the initial

syllables of the first six phrases of a hymn to Saint John the
Baptist:[25] *ut, re, mi, fa, sol, la.* The seventh note, *si* (sometimes *ti*), was added to the scale during the thirteenth century. These designations, with no change except that *ut* was
replaced by *do* during the seventeenth century, have continued to be useful in elementary music instruction and constitute a rare example of the survival of a pedagogic aid.

Several lesser musical theorists appeared elsewhere in
western Europe during the eleventh century and by their
writings gave further evidence of the awakening interest in
science. Among these may be mentioned Adelbold of Utrecht
(ca. 970–1026), Hermann the Lame (Hermannus Contractus)
(1013–54), Wilhelm von Hirsau (d. 1091), and Frutolf (this
name variously rendered; d. 1103). Each of these wrote one
or more tracts on music, and Hermann the Lame produced a
curious and interesting modification of Al-Kindī's notation
for the specification of pitch. None of them added anything
to the stature of acoustical science, however, and it may well
be, as Sarton suggests, that the very completeness of Ibn Sinā
had so sterilized the subjects he dealt with that nearly two
centuries would need to pass before he could be overtaken.

This was the time of Omar Khayyam [b. ca. 1038–48, d. ca.
1123], the Persian poet whose just distinction as a mathematician has been overshadowed by the romantic aura generated by Edward FitzGerald's famous paraphrase of his
Rubā'īyāt. Still farther to the east was occurring a contemporary technological development that was to be of far-reaching
significance. This was the invention, during the fifth decade
of the eleventh century, of the art of printing with movable
type. Credit for this invention goes to Pi Sheng, a Chinese
workman who first experimented successfully with type made
of fire-hardened clay.[26] The invention was subsequently (ca.
1314) improved by Wang Chen, who described both a species
of wooden type and a "lazy Susan" sort of revolving table to
store it within easy reach of the typesetter—a description that
has since been corroborated by the actual discovery in the

caves of Tun-huang of such a font of wooden type that can be reliably dated from about 1300. Specimens are also still extant of a later form of metal type that had been cast from type molds in Korea as early as 1403, and books printed on paper using such movable metal type can be dated as far back as 1409—at least three decades before the date conventionally assigned for the invention of the art of typography by Johann Gutenberg (1400?-68).

Another technological development of less moment but more acoustical interest is recorded in the eleventh-century treatise written by a little-known German monk and craftsman named Theophilus (also called Rugerus) (fl. ca. 1100). The style of his *Diversarum artium Schedula*, in its explicitness about practical techniques, is not unlike that of the *De Architectura* of Vitruvius. Theophilus was particularly concerned, however, with the handling and working of glass, gold, silver, copper, and iron, and he devoted several chapters to the construction of organs and organ pipes, and to the practical problems encountered in casting bells and cymbals. For both of the latter, he described in detail a "lost wax" method of preparing molds, and gave directions for insuring that a series of cymbals should fall in the proper tonal sequence by carefully weighing in advance the amount of wax used in making the mold. If this was not done accurately, however, he knew how to make a final correction for adjusting the pitch of the cymbal, which he explained by saying, "Should you wish the cymbal to be higher, you will file about the mouth underneath; but if flatter, round the rim in circumference."[27a] Theophilus also invoked an application of acoustics at the moment of crisis in the casting of a bell, when the molten metal is about to be poured into the mold: "In the meantime lie down, near the mouth of the mould," he instructed, "carefully remarking by listening how far within it may proceed."[27b]

The transfer of the stewardship of natural science to the West acquired more momentum during the twelfth century,

and became virtually complete by the close of the thirteenth
century. One channel of transmission more direct than most
was provided in the first half of the twelfth century by the
travels in Italy and Asia Minor of Adelard of Bath (England;
fl. 1116–42). The translation of Arabic mathematical treatises
engaged a substantial part of Adelard's attention, but he also
wrote on his return (ca. 1111–16) the brief but influential
Quaestiones Naturales. This was composed in the literary
style of a Platonic dialogue, with Adelard's nephew cast in
the role of questioner, but the subject matter more nearly
resembled that of the pseudo-Aristotle's *Problemata.* When
the physical nature of sound became in its turn the "natural
question," Adelard's patient explanations made it clear that
foreign travel had at least brought him into contact with the
Greek acoustical tradition as set forth by Boethius in his
Musica, to which he referred explicitly.

Adelard added something, perhaps, to the Boethian version
of spherical wave propagation when he pointed out that a
negative-pressure wave can also be transmitted in the same
way. As he put it, "just as when I form a sound by impelling
the air from me, . . . so when I draw the air inwards, and withdraw it violently from others, it is . . . perceived." The illusion
of understanding was quickly shattered, however, when
Adelard followed this up by "explaining the hidden truth"
that sound can be heard when "a solid wall is interposed between me and the hearers," because "every metallic body,
or even anything more solid than that, if such exists, is full of
porous interstices which afford a passage to so subtle a thing
as air."[28] Neither the meager novelty nor the muddy conclusion of this brief discussion of sound lends it very much
importance in its own right. The significant thing was that
some of the familiar old ideas about sound, and some of the
even older misconceptions about it, were indeed reestablished
and made available in western Europe during the early part of
the twelfth century.

Not long after Adelard's inoculation of England with Arabian science, intellectual borrowing received another westward impulse at the hands of the Spanish-Arabian Ibn-Rushd, better known as Averroës (1126-98). The star of Islam was already setting in the West and his patron, Al-Mansur, was the last of the Mohammedan caliphs to rule in Spain. With similar finality, Averroës was the last of the bright galaxy of great Muslim philosophers to be heard from. The tide of intolerance was rising and his intellectual liberalism earned him, before his death, both abuse and a period of internment under accusation of infidelity and heresy. His philosophy, which actually differed very little from Christian scholasticism, continued to be developed as "Averroism" long after his death, with the result that his influence was felt more acutely on later Jewish and Christian thought than it had been on Arabian.

Most of the Greek tradition preserved in Arabic and Syriac translations, as well as the Greek originals that still survived, were in the process of being rendered into the schoolmen's Latin during the twelfth and early thirteenth centuries by such translators[29] as Gherardo Cremonense (1114?-87), Dominicus Gondisalvi (Gundissalinus) (fl. ca. 1140), and Michael Scot (d. ca. 1235). The various commentaries on all the major works of Aristotle written by Averroës originally in Arabic were swept up in this tide of Latinization, and much of Averroës's influence stemmed from the acceleration his commentaries gave to the spread of Aristotelian doctrine. Although Averroës allowed himself for the most part to be confined topically to the things Aristotle had discussed, his amplifications would occasionally outrun his reference text. Thus, for example, in dealing with the palpability of colors, odors, and sound, Averroës argued that "sound is produced by a passion [*sic*! i.e., an effect produced on] of the air, but it is also [like odor] impeded by the winds; but yet it does not follow from this that it is a body."[30] This may not have advanced the understanding of odor very much, but it did tend to straighten out a few of the rambling notions advanced by even some of the post-Aristotelians concerning the corporeal entities of voice and sound.

Half a dozen men turned up in the West during the thirteenth century who were to exert a profound influence on the evolution of natural science. None of them can be said to have appreciably advanced the frontiers of the science of sound, and one must still look to the East for original contributions to acoustics during the thirteenth and fourteenth centuries. Nevertheless, the impact of these men on the growth of a central core of science from which acoustics would ultimately draw, demands that they be noticed.

Two were mathematicians: Leonardo (of Pisa) Fibonacci (1170–1250) and the German Dominican monk Jordanus Nemorarius (fl. ca. 1220). Neither left mathematical works that greatly outstripped those of their contemporaries in the East, but for the Latin West they represented the spark of mathematical originality that was finally to burst into flame in the late seventeenth and eighteenth centuries. Although he was less distinguished than Leonardo as a mathematical theorist, Jordanus earned his laurels as an applied mathematician by concerning himself with problems of mensuration and mechanics. The facts about his life are obscure, and it is not wholly certain to what extent his treatise *Scientia de Ponderibus* was improved and amplified by followers who used it, some with and some without acknowledgment. It is clear, however, that Jordanus revived in the West a scientific interest in mechanics, and that his work contained the seeds, if not the full flower, of such quantitative concepts as that of virtual work and turning moment.[31]

The Experimental Method Flexes Its Muscles

Robert Grosseteste (ca. 1168–1253), onetime chancellor of Oxford and bishop of Lincoln, deserves to rank as one of the prime movers in the rebirth of physical science in the West. He was remarkably free of superstitious fancies, and his integrity can perhaps be judged by the fact that after being made bishop of Lincoln he dared to quarrel with the pope and to refuse steadfastly to appoint a nephew of the pope to a minor church position, despite threats of excommunication. The study of natural philosophy that he introduced at Oxford was the first program of its kind anywhere to be based upon both mathematics *and* experiment. Grosseteste was primarily concerned with optics, whose laws, he believed, provided a rational basis for explaining all physical phenomena. This fixation led to strained logic sometimes, as when he said in a commentary on Aristotle's *Posterior Analytics*, "the substance of sound is light incorporated in the most subtle air."[32a] It is certain, however, that Grosseteste did understand the laws of perspective and that he was familiar with the magnifying properties of lenses, and it can be inferred that it was through him that this information passed to his most distinguished disciple, Roger Bacon.

Grosseteste regarded Nature's economy of effort (Aristotle's *lex parsimoniae*) as an objective physical principle and he based many of his rationalizations on its use. Thus he was led to invoke, as Heron had much earlier, the "shortest path" principle to explain the laws of reflection from plane mirrors, but his supporting argument was less lucid than the one advanced later by Witelo (Vitellonis) (b. ca. 1230-?).[32b] It is notable that the confusion of light and sound contained in his early commentary on the *Posterior Analytics* is entirely missing from Grosseteste's essay *De generatione sonorum*. Each of these works, however, contains a knowledgeable description of the elastic vibrations of a block or bar which sounds when it is struck, and each sets forth a notable anticipation of Poisson-ratio coupling between longitudinal and transverse strains that is not even hinted at in the original Aristotle. The passage goes:

An extension follows this movement of small parts necessarily as they move from their neutral positions, an extension of these parts along

a longitudinal direction and a contraction along the transverse direction; and in the return to the neutral position on the other hand
there occurs a shortening of the longitudinal dimension and an increase of the transverse dimension. And this motion . . . which follows the local motion of the tremor, is sound, or the speed natural to
sound. And when parts of the block shake, they move the air adjacent to them in a motion like theirs, and the motion travels to the
same sort of air contained within the ears and produces there a pressure on the body which is not concealed from the spirit, and there
results a sensation of hearing.[33]

The Dominican Albertus Magnus (1206–80) was a learned
scholastic who was almost as influential in fostering and
stimulating scientific investigation in Europe, at the University of Paris and at Cologne, as Grosseteste had been at
Oxford. In his writings, Albertus sought to combine the
teachings of Aristotle and of Ibn Sinā and Al-Fārābī, but he
was patently more concerned with the logic and tactics of
science than with the advancement of any of its branches,
and in due course he was overtaken even in this sphere of
influence by one of his students, Thomas Aquinas (1225–74).
As a consequence, the received state of acoustical knowledge
in western Europe during the thirteenth century advanced
very little beyond Aristotle—but it *was* received.

The experimental method found still another strong and
articulate advocate during the thirteenth century in the
Franciscan monk of Oxford, Roger Bacon (ca. 1214–92). At
the special invitation of Pope Clement IV, Bacon undertook
to summarize all that was then known of the physical sciences
in an *Opus Majus*, an *Opus Minus*, and an *Opus Tertium*, all
three completed in 1267. Unfortunately, his papal sponsor
died shortly after these works had been dispatched to him
and Bacon found little favor with Clement's successors. As a
result, when his *Compendium Studii Philosophiae* aroused
the bitter antagonism of churchmen a few years later, Bacon
was haled before a tribunal on trumped-up charges that suggested petty jealousies and was obliged to spend all but the
last of his remaining years in prison.

The *Opus Majus*, on which Bacon's fame largely rests, set forth with uncompromising objectivity that the study of natural science should rest solely on experiments, that such experiments should be planned, and that special apparatus should be constructed to carry them out. Such a conception of experimental science had its precursors in the practice if not in the theory of the ancients, and its expression had become progressively more articulate during the golden age of Muslim science. Even without these antecedents, however, the doctrine of Bacon's teachers and contemporaries would adequately support the viewpoint that his pronouncements were evolutionary rather than revolutionary. A good many "firsts" have been ascribed to Bacon from time to time, such as the invention of gunpowder, spectacles, the compass, and so on, but more recent studies have stripped him of most of these. There is credit enough due him, however, for his enduring contributions to the tactics of science. What Bacon had to say explicitly about sound occurs chiefly in his lesser known *Opus Tertium*[34] where, unfortunately, he does little more than demonstrate that his studies had included Boethius and Nichomachus as well as Aristotle.

While his contemporaries were advancing the cause of experimental science by precept, the physicist Petrus Peregrinus (sometimes Peter the Stranger; fl. ca. 1269) was powerfully forwarding the same cause by example. His *Epistola . . . de magnete*, completed in 1269, was a first and outstanding exemplar of the experimental method at work.[35] The first part of this treatise is devoted to his experiments, the second part to the construction of instruments based on the experiments. Petrus included a magnetostatic perpetual-motion motor among these instruments, "at the suggestion of others," according to Gilbert; but he can easily be forgiven for this in view of the fact that he also included descriptions of the two kinds of magnetic poles and their attractions and repulsions, the first pivoted compass needle, secondary magnetization by contact, and the action of magnets through

glass, water, and other intervening substances—prime results
that were not surpassed until Gilbert embraced them in his
own comprehensive treatise on the magnet in 1600.

It is noteworthy that the contributions of the foregoing
thirteenth-century scholastics and commentators displayed a
sharp contrast with the pattern established during the preceding six centuries, a pattern in which acoustical science had
flourished chiefly in the music of the Middle Ages. Much of
what was said about sound in the zealously Christian West
still adhered to the discussion of music, but the custom of
prefacing any writing about music with some consideration of
the "causes of sound" began to be paid only lip service—and
from Aristotle's lips at that. Enough has been said elsewhere
about the crippling influence of church strictures (see n. 4).
The enervating effect of reliance on even acoustical "authority" is made painfully obvious in the documentary evidence
from this era, which exhibits many strained efforts to ferret
out the "real meaning" of what little Aristotle and his commentators did have to say about the physical nature of sound.
But whether from such artificial or from natural causes,
there is clear evidence here of the growth of a phase in the
evolution of acoustics that was to reach its peak in the eighteenth and early nineteenth centuries—a phase in which the
leading men of science devoted their primary efforts toward
furthering the various other sciences that border and underlie
acoustics, leaving the synthesis of these disciplines in a science
of sound to be accomplished by those who follow.

In consonance with, and contributing to the identification
of the Western trend toward separation of the physics of
sound from its musical manifestations, two important treatises devoted to music as such appeared in Europe during the
early thirteenth century: the *Ars Cantus Mensurabilus* and a
Compendium Discantus. Both are often attributed to Franco
of Cologne (fl. end of twelfth century), but Combarieu suggests that the first was probably written by an earlier Franco

of Paris.[36a] These treatises were the earliest accounts of mensural music to appear in Christian Europe and they played an important part in the subsequent evolution of Western music. Among other things, each of them introduced a modified musical notation which resembles in form, if not in uniqueness of interpretation, the modern staff notation now in universal use.[36b] As a consequence, modern musical notation is surely "Franconian," whichever Franco is held to be its author; and the question is the less material because neither Franco should probably be credited with more than the effective transmission of these notions from their Muslim sources of origin (see n. 7).

Acoustical Climax in the East

Almost as if to demonstrate the extent to which the East had remained free from the curse of scholasticism, one distinguished contribution to music, and incidentally to acoustics, was made during the thirteenth century by Ṣafī al-Dīn (fl. mid-thirteenth century, d. 1294), the last and one of the greatest of the Arab musical theorists. In his *Kitāb al-Adwār* (Book of Musical Modes), probably written in 1252, Ṣafī al-Dīn proposed a melodic doctrine based on division of the octave into sixteen intervals, out of which could be formed a surprisingly large number of eight-note scales conforming closely to "just" intonation. Helmholtz called this "a tonal system very noteworthy in the history of the development of music,"[37] and it gave rise in due course to the "Systematist" school.

The usual introductory section devoted to the "bases of sound" was left out of the *Kitāb al-Adwār*, but Ṣafī al-Dīn repaired this omission in his *Risālat al-Sharafiyya* (The Sharafian Treatise on Musical Proportion), a more finished work written in 1267 for the Sharaf al-Dīn Hārūn. In the first discourse of this treatise, Ṣafī al-Dīn distinguished himself as the outstanding and virtually the only scholar before

Mersenne and Galileo to make any substantial advance beyond the physical understanding of sound revealed by the Greek scientists and their Roman commentators.

After reviewing the efforts of his predecessors to define the differences between sounds in general and musical tones, or "notes," Ṣafī al-Dīn concludes that a musical tone is a "sound for which one can measure the excess of gravity or of acuity with respect to another sound with which it is being compared,"[38a] which is not only a good operational definition but one that conforms almost exactly to modern terminology.

Al-Fārābī had clung tenaciously to the ancient confusion in which "firmness of the impact" was erroneously associated with the pitch of the resulting sound. In this connection, Ṣafī al-Dīn corrects his masters, saying apologetically, "It would be more correct to say that the sound is more intense, instead of more high pitched, when the impact is firmer." Elsewhere Ṣafī al-Dīn invokes devious arguments in the effort to justify some of Al-Fārābī's remarks, but this salvage operation is nowhere as successful as when he continues:

> it would have been possible, however, to reach this [Al-Fārābī's] conclusion by specifying that one had in mind the notes of wind instruments such as the reed flute (*yarā*); because by increasing the force of the impulse impressed [with the breath] one adds to the acuity [raises the pitch] of the sound heard from one and the same opening. It is for this reason, by the way, . . . one can produce with it [the *yarā*] more notes than it has openings, either by applying a more or less strong impulse [overblowing], or by closing certain of these paths [finger-holes] arranged for the air to escape.[38b]

Ṣafī al-Dīn added a few brief but highly suggestive remarks about voice production, which he explained by saying, "Here now is how the notes are produced in the throat of a man: the air applies a firm and violent impact to the cavities of the larynx; and this impact gives rise to a note."[38c] He then continued to deal further, in a qualitative way but more cogently than any previous writer had, with the phenomenon of resonance in an air column:

When in wind instruments air escapes abruptly and violently [as from the player's mouth, or from the opening of a reed], it collides with the tubular walls of these instruments, retraces its path, thrusts against the air that it meets and receives impulses in return, pushing back and being pushed back; spinning there in spiral fashion, from rebound to rebound under the force of the compressions and dilatations, it produces the notes, as they have said.[38d]

The "length and thickness" of a vibrating string had been mentioned together by Al-Fārābī in discussing the causes of "gravity," but he and all the others before him had failed to commit themselves with regard to the specific influence of thickness. Ṣafī al-Dīn took this hurdle in stride by saying that "greater length is a cause of gravity when one deals with a string; so is also more thickness and more relaxation of its tension." He did not achieve a mathematical statement of the relevant law of proportion, but he came close to doing so when he continued:

You will sometimes observe that the sonority of a relatively thin and short string is graver [lower pitched] than that of another which is thicker and longer. This can be caused by a greater tension of the longer string and by a greater relaxation of the shorter one. Nevertheless it remains true that a relatively thinner and shorter string lends itself better to high-pitched sounds.[38e]

A notable but anonymous commentary appeared in 1375, ostensibly addressed to the *Kitāb al-Adwār* of Ṣafī al-Dīn. It actually contained more than this, however, including comments on the sections devoted to the physics of sound in the Sharafian treatise and in the older writings of Ibn Sīnā and Al-Fārābī. In the preface he supplied for d'Erlanger's translation, Farmer ascribes the authorship of this commentary to Al-Jurjānī (1339–1413), and he says elsewhere, "there is no greater work than this masterly treatise except the *Kitāb al-Mūsīqī* of Al-Fārābī."[39] This rather overstates the case insofar as the physics of sound is concerned, and in this respect the work must be judged inferior to the Sharafian treatise. In dealing with the fact that "the masters of the art do not agree," Al-Jurjānī reverts to some of the Aristotelian locutions and pays rather more attention to Al-Fārābī and Ibn Sinā than to Ṣafī al-Dīn, to his own disadvantage. He finally does retain and reaffirm most of the conclusions reached by Ṣafī al-Dīn, although even these are rendered less convincing by the gloss of pseudo-sophisticated, but still spurious, efforts to rationalize the explanations he quotes from his reference texts.

On the other hand, Al-Jurjānī must be credited with breaking into wholly new territory when he touches on the problems of quantitative measurement. "One cannot measure quantitatively with the same ease the value of these causes," he says, "in order to determine the differences [of acuity and gravity]. . . . To date we have no means to measure the thickness or thinness of a string, nor to estimate its tension and relaxation. It is different for the length and shortness of strings. . . . In the case of wind instruments, we can estimate the width of the tube and of its openings; when they are circular, it is sufficient to measure their diameter. We can also measure the distance between these openings and the mouthpiece; but, to date, we do not have at our disposal any means for measuring the force or weakness of the breath."[40] It can be added, a little wryly, that his last lament comes uncomfortably close to remaining true even as of this twentieth-century date!

The onset of decadence suggested by the translation from the Sharafian treatise to Al-Jurjānī's commentary was symptomatic of the fact that the lamp of science which had been kept burning in the East had begun to flicker out during the thirteenth and fourteenth centuries as renaissance stirred in the West. The promise held out by the bold pioneering of Grosseteste and Bacon was still unrealized, however, and scientific inquiry made relatively little progress in either the

East or West during the fourteenth century. For example, Nicole Oresme (ca. 1323–82), best remembered as one of the founders of the modern French language, was a mathematician, economist, and commentator, but so far as physical acoustics is concerned, he had nothing new to offer. He suggested, as the early Greeks did, that sound may be nothing but motion, "and so perceived by hearing," but he added that this is "difficult to sustain and not relevant to the point at issue."[41] The points at issue with which Oresme was chiefly preoccupied were those concerned with magic and "things marvelous." He did hold that sound could carry a long distance in many ways and could be heard after its source had ceased to exist; but, alas, it was not reverberation that he meant, since he illustrated the latter point by referring to the continued ringing of a bell that had broken when it was struck.

The Renaissance Rampant

Of all the commentators on the ancient masters, the one whose words fall most quaintly, yet gracefully, on modern ears are those of Geoffrey Chaucer (1340–1400). How better could it be said that "air is broken . . . and rolls about with the fragments of sound,"[42] than as Chaucer sings it:

> "Soun is noght but air y-broken,
> And every speche that is spoken,
> Loud or privee, foul or fair,
> In his substaunce is but air;
> For as flaumbe is but lighted smoke,
> Right so soun is air y-broke."[43a]

As for the action of air on air in the forwarding of sound waves, and the Chrysippian analogy of expanding ripples on water, Chaucer put these ideas together this way:

> "And right thus every word, y-wis,
> That loude or privee spoken is,
> Moveth first an air aboute,
> And of this moving, out of doute

Another air anoon is meved,
As I have of the water preved,
That every cercle causeth other."
. . . .
"Now have I told, if thou have minde,
How speche or soun, of pure kinde,
Enclyned is upward to meve;
This, mayst thou fele, wel I preve."[43b]

In the late fifteenth century, when the Renaissance was burgeoning, the spirit of free inquiry came to full flower in Leonardo da Vinci (1452–1519), the illegitimate son of a dashing young Florentine lawyer. Leonardo was a companion of Savonarola and Machiavelli, an intimate friend of the arithmetician Fra Luca Pacioli, a pupil of Verocchio, and a contemporary and sometimes the rival of Michelangelo and Raphael. As the diversity of such a roster indicates, his interests were almost universal and he brought his unique talents to bear with lusty and contagious enthusiasm not only in the field of art but in almost every branch of natural philosophy. This eagerness led him to undertake many more inquiries than he could hope to carry through, but he did not hesitate to abandon the unfruitful ones when he could see that the results would not satisfy him. In his painting he worked slowly, and his perfectionism made him always reluctant to admit that any task was finished—but his paintings were masterpieces.

Leonardo started to write many books but he never finished any of them. His rough manuscripts he characterized as a "collection without order" of "strong and stern delight." Most were written with his left hand in backwards script that required mirror-reversal for easy reading, and they survived only as fragments that suffered wide dispersal after his death. These fragments later became prized acquisitions for museums and private collectors, and a compilation of them, in a translation of considerable distinction, has recently become available.[44] They reveal a remarkable intuitive grasp and qualitative command of all the natural sciences, and so clear a conception of the scientific method as to suggest that Leonardo's influence might have been pivotal in the evolution of science itself if only he had published his writings promptly. As it was, Leonardo's achievements were almost entirely first-personal and his influence on science and other scientists was scarcely felt in his own time.

In the field of mechanics, Leonardo expressed a good understanding of inertia, saying "every heavy body weighs in the line of its movement,"[44a] and recalling that "the thing which moves will be so much the more difficult to stop as it is of greater weight."[44b] He had a proper conception of turning moment and understood the resolution of forces and the relation of frictional resistance to weight and area of contact. His notes were frequently repetitious; Leonardo excused this characteristic plaintively by saying, "if I wished to avoid falling into this mistake, it would be necessary . . . that on every occasion . . . I should always read over all the preceding portion."[44c] Through such repetitions, however, he made it clear that he understood (perhaps as well as his translator did) the distinctions between impulse, force, work, and power; and in one instance he came close to disclosing the principle of conservation of momentum when he said, "If two balls strike together at a right angle, one will deviate more from its first course than the other in proportion as it is less than the other."[44d]

Leonardo anticipated Newton in declaring the equality of action and reaction, and in suggesting a universal attraction by which "every weight tends to fall towards the centre by the shortest way,"[44e] and "every part of an element separated from its mass desires to return to it by the shortest way."[44f] Leonardo explicitly posed the question whether heavy bodies fall more rapidly than light ones, and while he never quite offered an unambiguous answer, his preoccupation with the effect of air resistance on free fall suggests that his sophistication in this matter may have surpassed Galileo's. He did determine *by experiment* that free fall was a uniformly accelerated motion, that "anything which descends freely acquires fresh momentum at every stage of its movement[44g] . . . therefore at each doubled quantity of time the length of descent is doubled and also the swiftness of the movement."[44h]

With regard to acoustics, Leonardo shared with the ancients the knowledge that "there cannot be any sound when there is no movement or percussion of the air."[44i] He seems not to have leaned very heavily on the teaching of the ancients, however, if indeed he had been exposed to any; instead he made for himself careful studies of the propagation of waves on water and was led independently to regard these and sound waves as similar phenomena. Thus he recorded exuberantly: "The law of mechanics is the same in both instances! As waves upon water from the thrown stone, so do the waves of sounds go through the air, crossing one another without mingling, and preserving as their central point the place of origin of every sound. . . . There is but one sole law of mechanics in all the manifestations of force."[45] The principle of superposition that Leonardo declared in this way was to be restated in succession by Francis Bacon and by Huygens, and was eventually to be recognized as a basic feature of all wave phenomena in linear nondispersive media.

Leonardo also concluded from his observations of echoes that the wave motions of sound have a definite finite velocity of propagation, and he anticipated Galileo's discovery of sympathetic resonance by his observation that "the stroke of one bell is answered by a feeble quivering and ringing of another bell nearby; a string, sounding on a lute, compels to sound on another lute, nearby, a string of the same note" (see n. 45). Even modern sonar techniques for submarine detection have a historical antecedent in Leonardo's observation that "if you cause your ship to stop, and place the head of a long tube in the water, and place the other extremity to your ear, you will hear ships at a great distance from you."[44j]

In addition to being an artist and a natural scientist, Leonardo was a prolific inventor and an ambitious designer of military and civil works, and he did not hesitate to propose and plan the diversion of streams and rivers by canal systems or even the enlightened reconstruction of a whole city. He devoted himself with sustained passion to the planning of machines that might enable man to fly, but like many of his too elaborate proposals, these came to naught. He was always fascinated by the grotesque and the ugly, finding in these contraries a guide to perfection of form, and his sure knowledge of light and shade made works of art out of the notebook sketches he devoted to the study of anatomy. Leonardo attached himself in succession to many patrons, including finally King Francis I of France, in whose service he died; but he served first and always his own steadfast belief in the integrity of beauty and the validity of experimental observation.

There were no scientific disciples qualified to carry on Leonardo's scientific work immediately, but it was not long before his faith that "experience is never at fault"[44k] was ably seconded by Sir Francis Bacon (1571–1626), the Elizabethan philosopher whose judgment of sound as "one of the subtilest pieces of nature" has already been noted. In discussion of the origins of the so-called experimental age, Sir Francis Bacon is often confused with his predecessor Roger Bacon, a confusion easy to condone inasmuch as both talked a good bit about experimentation and neither did very much of it, Roger Bacon hardly any and Sir Francis only a few modest experiments dealing chiefly with heat phenomena. The interim growth of the art of printing and Francis Bacon's natural aggressiveness combined to give his remarks wider currency and greater influence, however, and the time was ripening for something to be done about it. Bacon dealt extensively with sound in the second and third "centuries" of his book on natural history (*Sylva Sylvarum*),[46] but he, too, made his principal contribution through a renewed endorsement of the "new approach" to scientific problems. His acoustical facts hardly went beyond those of Aristotle, from whom, indeed, he took most of them.

The Bridge from Mechanics to Acoustics

Before Bacon's libel of superficiality in the study of sound could be lifted, the house of mechanics itself had first to be put in order. Two other great men of Bacon's time were instrumental in bringing about the transition from Leonardo's inspired speculation to Galileo's quantitative logic. The first was Johann Kepler (1571–1630), the astronomer who is less renowned for the things he saw in the heavens than for the system and order he was able to discover in the mass of observational data carefully recorded and ultimately bequeathed to him by Tycho Brahe (1549–1601). Kepler published in 1618 a lengthy treatise called *De Harmonica Mundi* that contained a great deal of metaphysical musical nonsense about the harmony of the celestial spheres, but his closing peroration was preceded by a clean, simple statement of the third law of planetary motion! And this, together with the first two laws he had announced in 1609, was to provide Newton a century later with the key to the law of universal gravitation.

The definitive form which Kepler's laws gave to the kinematics of planetary motion was matched by the definitive form to which Simon Stevin (Stevinus) (1548–1620) had earlier brought the subject of statical mechanics. This Dutch mercantilist and military engineer wrote, in Flemish, a treatise on Statics and Hydrostatics (*Beghinselen der Weegkonst*, 1585), but this work received wider notice when it was collected with his other writings by Snel, translated into Latin, and republished as *Hypomnemata Mathematica* in 1608. The problems with which Stevin dealt[47]—the force triangle and the inclined plane, the hydrostatic "paradox," and the stability of floating bodies—were all topics about which Leonardo had entertained correct qualitative notions; but Stevin was now able to, and did, put them in explicit and quantitatively useful form.

It was another contemporary of Francis Bacon's, Galileo Galilei (1564–1642), who did more than anyone else before

Newton to bring the whole field of experimental science to at least the stage of adolescence. He not only made careful quantitative observations concerning the natural phenomena of everyday experience, as exemplified by the popular legend concerning his study of the period of oscillation of the hanging lamp in the cathedral of his native city of Pisa;[48] he was also among the first of the new and still-flourishing school of adherents to the Grossetestian doctrine of using deliberately planned experiments designed to confirm a tentative hypothesis. Whereas Stevin had concerned himself with the science of statics, Galileo pioneered in considering quantitatively the dynamical relations between force and motion. He developed an understanding of centrifugal force and was the first to show that the path of a projectile is a parabola. He made use of the concept of momentum, although he defined it as the product of velocity and *weight* rather than mass. This, and his treatment of the projectile problem, indicated that he understood clearly the dynamical relations that were to become Newton's first and second laws of motion, even though he formulated them in such a way as to be applicable only for gravitational forces.

Four publications of progressively increasing significance marked the successive phases of Galileo's life, and each brought him a storm of trouble. His experiments on falling bodies, performed while he was a professor of mathematics at Pisa (*Sermones de Motu Gravium*), led indirectly to his resignation from that post in 1591. His discovery of the four moons of Jupiter (1610) with his "Galilean" telescope, announced in the *Sidereus Nuntius*, led to his first appearance (1615) before the Court of the Inquisition and to the court's edict (1616) denying him further publication or espousal of Copernican heresy. His *Dialogo dei due massimi sistemi del mondo* was completed while he enjoyed the patronage of Cosimo de Medici II and was published in 1632, Galileo believing that the papal favor of Urban VIII had released him from the strictures of the edict of 1616.

Churchmen found Simplicio, the naïve character of these dialogues, too sharply drawn for their comfort, however, and Galileo, by then an old man of sixty-nine, was again haled before the inquisitors-general (1633). This time the court condemned him to imprisonment and demanded a public recantation, the circumstances of which gave rise to the legend that Galileo followed his denial of the earth's movement by murmuring sotto voce, "But it does move." It is now conceded that this incident is wholly apocryphal, but the story accurately reflects Galileo's stubborn integrity. He spent the first three years of his imprisonment writing his *Discorsi e Dimostrazioni Matematiche, intorno a due nuove scienze*, in which he drew together and summarized his major conclusions relating to the principles of mechanics. With publication denied to him in Italy, Galileo gave the manuscript of these dialogues to his Dutch friend (Louis [III]) Elzevir who brought it out at Leyden in 1638. Unfortunately, Galileo never saw the printed volume, as he became blind in 1637 and died five years later.

Galileo's early studies of pendular motion, as well as his modest skill in playing the lute, had led him to carry out many experiments dealing with the vibration of strings. Although some of these inquiries must have been conducted relatively early in his career, their description is chiefly to be found near the end of the first "day" of his 1638 dialogues, where he has Salviati say for him, "Impelled by your queries, I may give you some of my ideas concerning certain problems in music, a splendid subject."[49a] He then proceeds to explain the relation of pitch to frequency, consonance and dissonance and the frequency ratios corresponding to musical intervals, vibratory resonance, sympathetic vibrations, and the quantitative dependence of the frequency of vibration of a string on its length, diameter, density, and tension.[49b] Although none of these topics was new, a few of his explanations were, and the compact presentation of the whole subject in the form of objective conclusions demonstrably based on experiment was a highly significant contribution.

Galileo's identification of the *frequency* ratios corresponding to the musical intervals deserves further mention. The Pythagorean doctrine of simple numbers had been based on the relative *lengths* of two strings that would sound the two notes of a consonant pitch interval; and continued fixation of these length ratios had been fortified for centuries by universal experience with the fingering of stringed instruments. Galileo pointed out, however, that if the pitches sounded by two similar strings had been altered by changing the *tension* of one of them instead of its length, or by selecting strings of different *diameters*, the pitch would have been found to vary directly or inversely with the square root of these quantities, with the result that the ratio 4 would have become associated with the octave, 9/4 with the fifth, and so on. "In view of these facts," he has the interlocutor Sagredo say,

> I see no reason why those wise philosophers should adopt 2 rather than 4 as the ratio of the octave, . . . Since it is impossible to count the vibrations of a sounding string on account of its high frequency, I should still have been in doubt as to whether a string, emitting the upper octave, made twice as many vibrations in the same time as one giving the fundamental, had it not been for the following fact, namely, that at the instant when the tone jumps to the octave, the waves which constantly accompany the vibrating glass divide up into smaller ones which are precisely half as long as the former.[49c]

The beautiful experiment to which Sagredo refers made use of an ingenious method of generating surface waves on water and sound waves in air with the same source by partially immersing in a large vessel of water a goblet that could be made to "sing" either note of the octave when its rim was stroked.

A second experiment, less precise but equally ingenious, was addressed to the same problem. Galileo had noticed that when he scraped a chisel over the surface of a brass plate, it would sometimes chatter or "screech" with a musical (?)

tone; and that when this occurred there would appear on the surface of the plate a series of fine, parallel, regularly-spaced tool marks. By patiently scraping and listening, he

> also observed among the strings of the spinet two which were in unison with two of the tones produced by the aforesaid scraping; and among those which were separated most in pitch I found two which were separated by an interval of a perfect fifth. Upon measuring the distance between the markings produced by the two scrapings it was found that the space which contained 45 of one contained 30 of the other, which is precisely the ratio assigned to the fifth.[49d]

The Infancy of Experimental Acoustics

The broad influence exerted by Galileo on the general growth of theoretical mechanics has encouraged the assignment to him of credit for also launching the *science* of sound. This distinction needs to be shared, however, if not lodged entirely with the Franciscan (Minimite) friar, Marin Mersenne (1588–1648), who wrote more extensively on sound than Galileo did and who made many new and original contributions to both experimental and theoretical acoustics.

After devoting his early career to his holy order and to translating and commenting on classical texts, Mersenne published his first acoustical tract in 1627 under the title *Traité de l'harmonie universelle*. This treatise is, in part, a commentary in the classical tradition on the works of the classical commentators and contains relatively little of acoustical originality. Within the following decade, however, Mersenne wrote a handful, or more nearly an armful, of tracts on musical and acoustical subjects that were of prime importance. These began with two minor treatises titled *Préludes de l'harmonie* and *Questions harmoniques*, each of which bore the date 1634; and these were followed by one of his major works, the *Harmonicorum libri*, of which some exemplars are dated 1635, some 1636. This work was in eight *libri* (books), although the number "VIII" did not appear in its title, and it was often bound in the same volume with another major work, the *Harmonicorum instrumentorium libri* IV, also dated 1636. This pair of tracts was reprinted from the same plates some time later; and after being supplied with a new title page, a second renewed dedication, and sometimes with a four-page interpolation entitled *Liber novus praelusorius*, the whole was reissued as an *editio aucta* under the title *Harmonicorum libri XII*[50] in 1648, the year of Mersenne's death. Still another set of Latin tracts dealing chiefly with aspects of mechanics was published in 1644 as *Cogitata Physico-mathematica*;[51] and although one of these bore the ubiquitous subtitle *Harmoniae*, its "propositions" allocated to vibrating strings and to the speed of sound consisted mostly of restatements of earlier results rather than accounts of new experiments.

In the meantime, Mersenne's great encyclopedic *Harmonie universelle* had also appeared in 1636. But both it and his contemporary Latin books on "harmonics" bore prefatory authorizations dated 1629, suggesting that the king's council as well as Mersenne's ecclesiastical superiors had seen at least part of these manuscripts well in advance of their nominal 1636 dates. It can be safely inferred from such internal evidence that the period of Mersenne's greatest acoustical productivity is spanned by the dates 1627–38, but it is hardly possible without further study of his voluminous correspondence to assign dates to the individual tracts with any certainty, or to resolve in any greater detail the exact sequence in which he performed his various experiments.

The *Harmonie universelle*[52] would be a fascinating book if only for the puzzling contradictions it presents; but it also contains a wealth of original material and furnishes a revealing glimpse of the growth of sophistication in the physical sciences during a critical period of their adolescence. Its nineteen *livres* were marshaled by the printer under six separate paging sequences, "contrary to my intent," as Mersenne insists in the general preface by way of explaining to the reader why he will not supply an index. Fortunately, he relented before the work was actually published and did supply such an index, in which the six paging sequences were identified serially by the letters *A* to *F*. The most important function this index serves is to establish Mersenne's definition of the contents and makeup of the *Harmonie universelle*, which, as Brunet testifies, "is rarely found complete. The pieces which compose the two volumes [but often in one] are not always placed in the same order."[53]

Parts of the *Harmonie universelle* are also alleged by Brunet to be French translations from the Latin *Harmonicorum libri*, but if there is any such linguistic coupling it must be judged to be very loose. The first four books "des Instruments" of section D do resemble the *Harmonicorum instrumentorium libri IV* of 1636, and the first book "de la nature & des proprietez du Son" of section A and the book "*de la Voix*" of section B contain parts that resemble the *Traité de l'harmonie universelle* of 1627; but it is just as pertinent to remark that many of the "propositions" of the Latin *Harmonicorum libri* read like condensed summaries of corresponding propositions appearing in the French.

In any case, the French version is more discursive by far than the Latin, by nearly four-to-one on the whole, and it contains a good bit of material that does not appear anywhere else in Mersenne's writings. The first book of section A might be expected to be the earliest but any attempt to date it before 1629 is challenged by the fact that it includes references to Galileo's *Dialogo* of 1632 and to Mersenne's own two treatises of 1634. It does contain more that is naïve, however, than do those which follow, as Mersenne himself recognized when he said, "It is necessary to point out that there are many things in the first book that need to be modified according to what is in the third book, and according to the experiments that everyone can perform at his leisure."[54a]

The uncertainties of dating and derivation suggested by the foregoing are compounded by Mersenne's redundancy and by his eager use of proportion to extrapolate his observations and to proliferate numerical examples. Thus, when the same basic experiments turn up in one after another of Mersenne's tracts with different numbers in evidence, it is not always clear which is the original experiment and which the illustrative example; unless, that is, one is willing to make the tacit assumption that experiments incline always toward improved precision and toward firmer conviction on the part of the experimenter.

A typical case in point is furnished by the method of measuring the speed of sound by timing the interval between seeing the flash and hearing the report of guns fired at a known distance. The principle of this method had been described a century earlier by Francis Bacon in one of his proposed, but not performed, experiments designed "to try exactly the time wherein the sound is delated [carried]." Thus, Bacon continued, "let a man stand in a steeple, and have with him a taper; and let some veil be put before the taper; and let another man stand in the field a mile off. Then let him in the steeple strike the bell, and in the same instant withdraw the veil; and so let him in the field tell by his pulse what distance of time there is between the light seen and the sound heard: for it is certain that the delation of light is in an instant. This may be tried in far greater distances, allowing greater lights and sounds."[55]

Mersenne took up this problem by first remarking, as the ancients had, that "sound can fill the sphere of its activity only in a space of time"; but what he proposed to do about it had a brand new flavor of arithmetic detail. "It will be necessary," he said, "to make several experiments to know if the delay of the sound varies with the separation; for example, if a sound made 2000 paces away will be heard only two second minutes after it has been made, and if it will maintain always the same proportion in its retardation."[54b] Note in passing that the term "second minutes" and such variants as "second minutes of an hour" occur often in early seventeenth-century texts. Each refers to the time unit derived by making successively a first and then a second "minute" (= small = 1/60) division of the hour, and hence to the ordinary second.

The problem of timing was one to which Mersenne eventually devoted much attention, but at this stage he merely pointed out that the pulse beat is more reliable for timing than respiration, "which carries the disadvantage of being voluntary." He followed up these conclusions about the utility of the pulse as an interval timer with an engaging sample of extrapolation:

> Suppose then, for example, that the natural well-tempered pulse beats 3 times before one can hear a sound made 500 paces away . . . and that there are 66 beats of such a pulse in a minute of an hour. I say then that the pulse beats at least 18 times before one hears the sound of a canon, an *arquebuse*, a trumpet, a bell, a *marteau*, a *tonnerre*, or any other instrument at distance of one of our leagues; and consequently that a sound which is strong enough to be heard around the world could only be heard after a time in which the pulse would beat 129,600 times, . . . but a sound would not last so long nor be strong enough to be heard so far away, unless God wished to produce such a sound; that will perhaps be when the Angels sound the Trumpet on the great day of Judgment to summon those who are about to die.[54c]

The speed of sound on which the first of these examples seems to be based is a good bit too high (unless *pas* [pace] has been misprinted for *pied* [foot]), and the speed is almost as much too low in the second example. It was not long, however, until Mersenne's experiences converged on the judgment that, "First, it is certain that sound goes through the air with the same speed be it strong or feeble, and of whatever kind, for example that of the voice or of a pistol or musket, & c., and be the wind contrary or favoring, as we have tested many times very exactly. In the second place, that it goes 230 toises [1 toise = 6 Paris, or "Royal," feet (*pied de Roy*)] in the time of a second minute, as we likewise observed on high mountains as well as on the paths in the park of M. Verderonne, and elsewhere."[54d,56a]

With this revised value of the speed in hand, Mersenne lost no time in shortening to 21 hours 5 2/3 minutes (!) his previous estimate of the travel time for a round-the-world sound wave;[54d, 57] and in the *Cogitata* he used the same revised speed to correct his estimate of the delay in hearing the trumpet of the Last Day, by observing that "it will be perceived everywhere on the earth in about ten hours from the point at which it sounds."[56b]

In spite of the fanciful climaxes of these passages, it remains to Mersenne's credit that he was tackling with serious purpose the problem of deducing the numerical value of an important physical constant, and doing so with very creditable results in view of the rudimentary state of mensuration in general and the near absence of reference standards applicable to the measurement of either short distances or short time intervals. With regard to the latter, the extensive and distinguished literature dealing with the history of clocks supplies ample evidence that the big "breakthrough" in the perpetual quest for ever more accurate time-keepers came with the adoption of the pendulum, and eventually its circular analog, the balance wheel, as a speed regulator for mechanical clocks. Galileo's name is traditionally and correctly associated with the discovery of the quasi-isochronism of the simple pendulum and of the law of variation of its period with length. But in this case, as in most others, Galileo expressed his conclusions almost exclusively in terms of ratios. This may have served his reputation well, inasmuch as when he departed from this practice, as he did in estimating the acceleration due to gravity in free fall, his numerical results were grossly inaccurate. On the other hand, Mersenne's approach to the pendulum problem was just as typically numerical, and he appears thus to have been the first to make a direct determination of the period of a pendulum.

Just when Mersenne first carried out this experiment is still obscure, although it must certainly have been at an early stage in the preparation of the 1636 manuscripts, since there are repeated references throughout both the French and Latin versions to a "chorde [string] three and one half feet long each of whose returns [*retours* = swings or vibrations] will last exactly one second minute." And although Mersenne did occasionally make use of his pulse for timing, he soon abandoned the then-common assumption that one pulse beat corresponded exactly to one second, as evidenced by his reference to a pulse rate of 66 per minute in the example cited above, and by his recommendations that a pendulum of adjustable length could be used by doctors to measure variations in the pulse rate and "how the passions of cholera and other fevers hasten it or retard it."[54e]

He also suggested that astronomers could use such an adjustable "chorde" to measure eclipses of the sun and moon, and that musicians could thereby "make known to everyone the time to be given to each measure in singing all kinds of music." In fact, Mersenne concludes, such a "chorde can serve all the uses that one makes of ordinary clocks [*horologes*], which it surpasses in accuracy. Besides which, as experience teaches, for two half-sous one can make three or four clocks which will mark second minutes, by attaching a chorde three and a half feet long to a peg, for if one attaches to the other end something heavy which hangs freely toward the center of the earth, each of its returns will last exactly one second minute."[54f]

Mersenne did not declare unequivocally what primary standard of time he used, but it seems safe to presume that he relied on the time for the earth to turn through some arbitrary angle. It was common practice in his time to correct the rate of mechanical clocks by daily, or even more frequent, reference to the sundial. Certainly he could have used this method inasmuch as he noted that "the goings and comings [of his 3 1/2 foot pendulum] would last at least a half hour,"[54g] which is a time interval just barely long enough to be "read" from a sundial without intolerable errors. Mersenne also made use of such watches and clocks as were available, although he seemed to place more reliance on the accuracy of his pendulum: "for example, the same common clock, of

which I have often measured the entire hour with 3600 returns of the three and a half foot chorde, does not at other
times make its hour so long, for then it is only necessary to
make the chorde three feet long to have 900 returns in a
quarter of an hour indicated by the clock; and I have experimented on a watch with wheels made expressly to mark the
single second minutes, for which a chorde of 2 feet and a half
or thereabouts serves to make turns equal to the indicated
seconds."[54h]

Although it remains uncertain how Mersenne referred the
period of his pendulum to the *prime mobile*, there was no
doubt about the method used a few years later in experiments
conducted at Bologna by the Jesuit father Giambattista Riccioli (1598–1671). His objective was to make a pendulum that
would beat seconds *exactly*, and to test his success he counted
the total number of swings it made during time intervals that
were long enough to be measured with some accuracy in
terms of the rotation of the earth. The pendulum was kept in
motion by giving its hanging weight another gentle push every
few minutes as required to keep the amplitude of its swings
about the same. Four pendulums of slightly different
"heights" were tested: the first two gave 21,706 and 87,998
swings for intervals of 6 and 24 hours as measured by a sundial; the other pair yielded 86,998 swings for another 24-
hour interval, and 3,212 twice and 3,214 once for a shorter
interval, as measured by the passage of fixed stars through
the meridian.[58] Nine other patient Jesuit fathers worked in
relays to help make the long counts, and one can only con-
template with awe how one of them would have felt if he had
mixed up the count during the twenty-third hour! As the
counts themselves indicate, Riccioli did not quite succeed in
making a pendulum with exactly a one-second period, but his
experiments certainly deserve to rank as a heroic effort in the
quest for precision.

Mersenne's interest in the vibration of musical strings was
closely related to his studies of the pendulum and probably

preceded them chronologically. Galileo's *Discorsi*, in which he summarized his analysis of vibrating strings, appeared two years after Mersenne's 1636 publications, but it is reasonably certain that Galileo's work on strings preceded Mersenne's. It is equally certain, however, that Mersenne attacked this problem from a fresh viewpoint, as evidenced by the fact that his experimental approach was entirely novel and introduced into the field of mechanics for the first time the quantitative use of scale-model experiments.

Mersenne began by summarizing the rules concerning the effects of length, tension, thickness, and density on the frequency of vibration of a string. He added something to this familiar account, however, by specifying that when the octave is secured by quadrupling the weight with which the string is stretched, this larger weight needs to be augmented by 1/16 of its value in order to make the octave true.[54i, 57b] The need for such a second-order correction of the simple rules of proportion might be laid to systematic errors of measurement, and Mersenne offered no explanation of it; but it is more suggestive to observe that this is qualitatively just the kind and size of correction needed to compensate for the finite stiffness of a real, as contrasted with an ideal, vibrating string.

In the passages just cited, Mersenne followed the usual practice of stating the rules for strings in terms of ratios, but he made his most important contribution when he went further and sought to determine the actual number of vibrations per second made by a string when it sounded in unison with a particular musical note. The scheme he devised for doing this involved scaling down the frequency by working with very long strings that would vibrate slowly enough to "allow leisure to count the vibrations."[54j] Then, as Mersenne explained:

> there is no difficulty in finding the number of returns for each string proposed, for if one extends to 10 or 12 toises the length of a monochord . . . one can easily count its returns, the more so if it

makes a very small number, for example 2 or 3 in each second. But one needs two or three [observers] to note exactly the number of these returns, one to count the returns while the other counts the seconds, whence if one divides the number of seconds into the number of returns, one will know how many it makes in each second. And if one extends a string from a spinet or a lute to 100 or 120 feet, as I have done, one will find that each return of that string occurs in a second, and that half the same string makes two vibrations in a second, that a quarter of it makes 4, the eighth part 8, the sixteenth 16, the thirty-second 32, and so on.[54k]

It is clear that Mersenne could not have verified by direct counting the last few steps of subdivision he cites because, as he notes elsewhere, "ten vibrations [per second] are the limit of image formation, or of counting by eye, as will be clear to anyone making the experiment."[56c, 54m] The first two or three stages of division were accessible, however, and Mersenne verified in this direct way that the actual frequency of vibration of his oversize "string" varied inversely with its length. In other trials he verified also the usual rules of proportion for tension and diameter as well as length, using in one case a hemp rope 90 feet long and one line [= 1/12 inch] in diameter, and in another a brass "string" 138 feet long and 1/4 line in diameter.[57c]

This was one of Mersenne's most often repeated experiments and descriptions of it recur in all his works, with the usual diversity of exemplary numbers in evidence. Tests involving a bronze string 15 feet long and gut strings of 15, 17 1/2, and 18 feet in length were used to determine the frequency of selected musical notes by finding what shorter segments of the long string, under the same tension, would sound in unison with the reference note. Mersenne described the problem by first pointing out that

since nature has not set up for us any sound which we may take as a fixed norm for the rest, just as [it has] not [set up any] measure and weight by which we may weigh other quantities and weights, we must use something artificial. Let us therefore take the pipe of an organ, the sound of which seems to be more uniform than other sounds, . . . Now let the height of this pipe be one Royal foot, or to use the lowest-pitched organ pipe that I have on hand, let it be 11 3/4 inches, and its width or diameter 1 1/2 inches. It should be noted, moreover, that this pipe is closed, what the French call a *tuyau bouche.*

Lest, however, anyone be too dependent on this pipe, I make in unison with it a brass chord whose length, thickness, tension, and weight I shall specify: the diameter then of this chord is 1/4 line; its length 3/4 of a foot, that is 9 inches; and the weight of the chord is 8 grains approximately; finally, its tension is made by six pounds and ten ounces or 5/8 of a pound. Now in order that the number of recursions ["returns" = complete vibrations] of that chord may be perceived by sense itself [i.e., counted directly], a chord of fifteen Royal feet will have to be stretched with 6 5/8 pounds, and since it contains a chord of nine inches twenty times, this chord of nine inches will have its recursions twenty times in number the recursions of the chord fifteen feet long. Since then it is determined by experience that a chord stretched with 6 5/8 pounds and fifteen feet long makes 10 recursions in the space of a second, it necessarily follows that that same chord nine inches, or 3/4 of a Royal foot long (as it should be to be in unison with the aforesaid pipe), has 150 [should be 200; Mersenne corrected the slip in the *corrigenda* of the 1648 edition and when he restated this example in the *Cogitata*[55d]] recursions in the same space of time. But if it be made an octave higher, it will have 300 [400], and so on."[57d]

In Mersenne's French version of the analogous experiment, he took the *G re sol* [second *G* below middle-*C*] of a four-foot open organ pipe as his *ton de Chappelle*, and matched this with the vibration of a ten-inch segment of a 17 1/2-foot string. This tone, "which is about as low as my voice can descend," Mersenne stated explicitly, returns 84 times per second; but the numbers he used in describing the model experiment, after correcting an obvious slip in which *four* is written for *two*, lead to a frequency of just 80 cycles per second for this *ton de Chappelle.*[54j]

In still another version of this classic experiment in which the 18-foot string appears, Mersenne said, "it is clear from observations frequently repeated that a string vibrates a hundred and four times a second when it is in tune with a two-foot

closed organ pipe, which produces an interval of a fourth
above my lowest voice. This pipe the organists call C *fa ut*
[an octave below middle-C]."[56d] The quantitative interpretation of this example is not free from ambiguity, but if it be
presumed that the reference tone in question was that of a
"C-pipe" from a "two-foot closed" organ stop, rather than a
closed pipe exactly two feet long, then the G a fourth lower
would give a frequency of 78 Hertz (Hz) for Mersenne's
"lowest voice," in reasonably fair agreement with the preceding example.

If further evidence were needed, these examples would
show that Mersenne was no slave to the numbers he used. On
the contrary, he dealt with them casually, almost carelessly,
not even bothering to comment on the tolerance range of
"100 to 120 feet" for the length of the extended spinet string
for which "each return occurs in a second," or on the spread
between 80 and 84 Hertz for his *ton de Chappelle*. Just establishing the existence of such numerical relations and their
validity as a description of physical phenomena was the
novelty that Mersenne cherished, rather than the kind of
precision sought by Riccioli. Thus, he treated it as relatively
incidental when he decided that his "seconds" were a little
too long, and that his "chorde of 3 1/2 feet" might better be
one of just three feet.[56e] And no qualms about absolute accuracy dampened in the least the enthusiasm with which he
urged that music could be played or sung in the same way by
everybody in the world, if the composer would just indicate
his intentions by annexing to the score one number giving the
reference frequency for pitch and another number giving the
length of a pendulum whose swings would indicate the time
to be allotted to each measure.

As for the absolute value of Mersenne's reference frequency
for pitch, hardly any conclusion can be drawn about it that is
not contradicted by Mersenne himself. The 104 Hz he assigned to C *fa ut* and the 80 and 84 for G *re sol* would correspond to frequencies of about 347, 356, and 373 Hz when

referred to the treble A ordinarily used as the reference pitch for tuning—figures at or below the lower limit found by Ellis in his definitive study of the evolution of musical pitch.[59] Elsewhere Mersenne referred to the *ton de Chappelle* being made with 60 returns in a second, which is a major fourth lower than the 80 Hz given above. A few pages farther along he gave 48 Hz for the frequency of a four-foot open organ pipe, which is ridiculously low; yet he constructed on this base a table of pitch numbers[54n] whose treble A at 480 Hz has sometimes been taken as typical of the "chamber pitch" of Mersenne's time!

The irresponsible abandon of these numbers does Mersenne no credit, even though it reflects accurately the chaotic state of standardization in musical pitch, or rather the lack of it, both in that period and for the next two centuries. Nevertheless, Mersenne's principles of standardization were sound, even though his own practice was variable and his measurements unreliable. Euler renewed a century later his suggestion that the calculable frequency of vibration of musical strings could furnish a convenient and reproducible standard of pitch.[60] But even these potentially useful proposals could not prevent the A's of church organs, and of the pitch pipes and tuning forks used as pitch references by musical organizations and performers, from scattering between 275 and nearly 500 Hz throughout the eighteenth and nineteenth centuries.[59] The median pitch trended generally upward, but its rate of rise was checked in the neighborhood of 435 Hz in the middle of the nineteenth century, and it is now standardized internationally at $A = 440$ Hertz. Of course, strong incentives for standardization were furnished by increasing demands on the compatibility of local and transient instruments of relatively fixed tuning. It is suggestive to notice that the frequency spread in reference pitch during these last few centuries has varied inversely with the growth of transportation and communication. Perhaps reference pitch owes much of its present stability to the engineers who have shrunk the world with air travel and electronics.

Another vehicle for Mersenne's experimental ingenuity was furnished by establishing a second front, or what may actually have been his first front, for attack on the problem of measuring the speed of sound. The scheme for doing this required merely that the delay be measured between the emission of a sound and the return of its echo from a reflecting surface at a known distance. The method he used for timing the echo delay, however, showed the true flash of experimental genius, and its combination with his urge to convert experience into quantitative measurement was typical of Mersenne at his best.

Mersenne began these studies by considering how far one must stand from a reflecting "body" in order to hear an echo. Polysyllabic words and phrases at once became test objects. If these were spoken too close to the reflector, the echo of the first-uttered syllables "would return too swiftly to the hearing and confound themselves in their encounter with the others. I have nevertheless found experimentally [*experimente*]," Mersenne continued

that Echo returns one syllable at 22 geometric paces [22 × 5 = 110 Royal feet], but one can still make several trials [*expériences*] to shorten this distance. As for an Echo which returns 2, 3, 4, & c. syllables [in the clear, that is without overlapping by the echo], it is necessary that it [the reflector] be 2, 3, or 4 times more remote. . . . one can deduce what the speed of sound is by [such] trials made with Echos, inasmuch as one easily pronounces two syllables one after the other, of which the Echo is heard while the pulse beats once, that is to say in the time of a second minute."[540]

These were capital ideas but very bad measurements, both of syllable timing and of the minimum distance for an echo. It is their crudity, in fact, and the mention of only the pulse beat for timing, that supports the suggestion that these were among Mersenne's first acoustical experiments, as does also

the fact that their description occurs in the first book of
Harmonie Universelle, which Mersenne felt obliged to warn
the reader not to take at face value without checking the third
book.[54a] His technique soon improved, however, and by the
time this experiment turned up again in the third book, he
had adopted the pendulum for timing: "one will easily prove
with an *horloge à secondes minutes* [seconds pendulum] . . .
that it marks one second minute for the pronunciation of
seven syllables by its first swing [*tour*] and the reverberation
of the Echo by its return [*retour*]." Then, after commenting
on the constancy of the swiftness of Echo in any wind and
weather, he continued, "It appears one can deduce the speed
of the voice and of other noises by means of Echo, for inasmuch as it returns the seven syllables *Benedicam Dominum*,
or such others as one may wish [to use], and returns them in
one second minute, the last syllable -*num* makes 485 Royal
feet in going, & as much in returning, in the time of one second, that is to say 162 toises or thereabouts."[54p]

Mersenne was not content to stop with these bare bones of
a measurement. The phenomenon of echo had received ample
notice in classical literature, and he took it on himself to
review the plausibility of some of the accounts and to "de-
bunk" the more extravagant of them, such as the report of an
echo that would reply in Spanish when spoken to in French.
Mersenne stoutly denied this possibility, although he almost
convinced himself that one could devise a special series of
sounds whose echo might lead a listener to *think* he had
heard the response in a different language.[54q] He was on
firmer ground when he disposed of the claim that the echo
from a tower near the Aventine Hill in Rome could repeat
the entire first verse of the *Aeneid* eight times. On the basis
of his measurements of syllabic rates, he estimated that these
eight repetitions would require at least 32 seconds [try it!],
that the reflecting surface would need to be half a league
distant, that there would be a confusing lot of other reflec-

tions heard before one would return from such a distance, and that anyway no voice could be strong enough to be heard at that distance![54r]

The linguistic pace represented by speaking the seven syllables of *Benedicam Dominum* in one second is relatively fast, as any reader can quickly convince himself by trying to repeat it rapidly while watching the sweep-second hand of a watch. And to declaim the phrase in *exactly* one second, with enough reliability and repeatability to allow the syllable structure of the phrase to serve as a split-second timer, must have required much patient practice; perhaps Mersenne's training in his holy order stood him in good stead here. In any case, his numerical result for the speed of sound was much closer to "right" than any cautious estimate of probable error would lead one to expect. If it is assumed, with Koyré,[58] that Mersenne's "Royal foot" had the length 0.3287 of a meter in modern measure, then the one-second 970-Royal-foot round trip for *-num* would correspond to a sound speed of 319 meters (or 1,046 modern English feet) per second, which is less than 10 percent below the accepted value for the speed of sound in air at ordinary temperatures.

One might expect that greater precision could have been obtained if Mersenne had repeated the test phrase continuously in cadence, and if he had then varied the distance from the reflecting wall until a position was found where the returning echo would accompany the succeeding phrase exactly in unison. Thus, matching the whole transmission for coincidence instead of relying on just the sequence of its terminations is essentially the modification of the experiment introduced by a young Oxford fellow, Joshua Walker (1655–1705), a half-century later. Walker produced a *periodic* sequence of short sharp sounds by clapping two thin boards together in synchronism with the swings of a half-second pendulum, whereupon he found that time coincidence between each clap and the echo of a preceding one could be determined with such precision that "the place could easily be found within a few yards."[61]

As a matter of fact, the discrimination capabilities of the ear are such as to suggest that in a carefully conducted experiment of this kind, Walker should have been able to find the "place" within a few feet or less, rather than "within a few yards." Hence it may not be surprising that the error in his result was twice as big as Mersenne's. What is surprising is that, in spite of its merit, this method of sound-speed determination was neglected thereafter until the mid-nineteenth century when Gustav Emil Kahl (1827–93) revived it,[62] retaining Walker's technique of folding back the transmission path by using reflection. About a decade earlier, Johannes Bosscha (1831–1911) had picked up the method and modified it in still another way by substituting for the echo the sound produced at a known distance by an electrically operated hammer,[63] a modification that Karl Rudolph Koenig (1832–1901) was quick to adopt for the apparatus he devised for the measurement of sound speed in 1862.[64]

In due course, which in this case means after another half-century of independent collateral development in the field of instrumentation, it became feasible to substitute for the observations of the coincidence of short sound pulses what amounted to observations of the coincidence of single cycles of the two continuous wave trains comprising a simple standing wave system. This modification made its first appearance in modern electronic trappings as the Pierce ultrasonic interferometer,[65] in which the reaction of a standing wave system on an oscillating quartz crystal used as a sound source was observed to go through measurable cyclic changes each time the distance between the source and a reflector was changed by a half wave-length.

Mersenne never succeeded in resolving to his satisfaction the discrepancy between his two major experiments on the speed of sound, which stubbornly continued to yield 162 toises per second for the speed as determined by echo timing and 230 toises per second as determined by his experiments on gun-blast timing. In retrospect it seems clear that the procedures involved in the echo experiments were inherently better insulated against errors introduced by personal reaction times, although the absolute accuracy of this method was crucially dependent on the highly individual skill with which he could repeat a spoken phrase with accurate timing. In the method of gun-blast timing, on the other hand, the observer could not control the initiation of the blast, and hence there was usually an opportunity for one, two, or three "reaction times" to be involved in the timing, depending on the details of procedure; and the absolute accuracy was further jeopardized by the difficulty of selecting an unobstructed course of adequate length and of measuring its length with accuracy.

Nevertheless, the blast-timing experiments had the compelling appeal of directness, and nearly all the experimenters who undertook to measure the speed of sound during the next two centuries placed their major reliance on this method. Mersenne himself, without the benefit of more sophisticated hindsight, led in this placement of reliance, if one can judge by the fact that in all his later tracts he gave for the speed of sound only the one value 230 toises per second. This figure corresponds to 1,380 Royal, or "Paris," feet per second, which is approximately 1,471 feet, or 448 meters, per second in modern measure (based on the conversion factors, 1 toise = 6.3945 modern English feet = 1.949 meters). At least two other numerical values for the speed of the sounds of gunfire have been ascribed to Mersenne by later experimenters, and all three values have then been propagated from one to another of the subsequent historical accounts and tabulations. Lenihan has shown convincingly, however, that all the other "values attributed to him [Mersenne] have been obtained by expressing this figure [230 toises] in different units, or in the same units with slightly different conversion factors."[66]

One of Mersenne's contemporaries, Pierre Gassendi (1592–1655), has been credited by many writers with having been the first to repeat the measurement of sound speed by blast timing, and some even give him priority over Mersenne, but Lenihan has also shown that these are spurious attributions and that there is no evidence that Gassendi ever made an independent measurement of sound speed (see n. 66). It is true that Gassendi was keenly interested in dispelling the Aristotelian notion that the speed of sound depended on the strength of its source, if only because in his philosophical writings he had followed Galileo in the revolt from Aristotelian authority. For this reason he must have welcomed the result of the experiment he described as follows:

> It is a matter of experience that sounds small or large, made at the same place, are carried in the same time to the place where they are heard; this can easily be observed from the sounds of artillery heard over a distance of two or three miles if, having observed the instant when the flash is produced, one counts the pulse beats or the oscillations of a pendulum until the sound arrives at the ear; one finds that the oscillations, which are of course of equal duration, are of equal number whether the sound is made by a large weapon, such as a cannon, or by a small weapon, such as a musket.[67]

Gassendi had already referred to Mersenne's measurement of sound speed and did not himself offer any numerical value other than the one he quoted from Mersenne. Moreover, the similarity of style in the remarks quoted and in parallel passages from Mersenne's work suggests that he may have been merely echoing Mersenne's conclusions. On the other hand, Gassendi *could* easily have performed the simple qualitative timing experiment he described without either knowing the period of his pendulum or bothering with the distance measurements needed for computing the speed, and it can be inferred that he probably did satisfy himself in this way that the travel times were the same for large and small sounds.

Rise of the Scientific Academies

In 1657, shortly after the death of Mersenne and Gassendi, the first of the post-Galilean academies was established in

Florence as the Accademia del Cimento. The nucleus of this novel organization was the entourage maintained by the grand duke of Tuscany, Ferdinand II (1610-70), eldest son of Cosimo II (1590-1620) and an enthusiastic amateur and patron of science. The protection extended to Galileo earlier by Cosimo II had been continued by Ferdinand II, although not effectively enough to rescue Galileo from the rigors of the Inquisition. Ferdinand's younger brother, Leopold de Medici (1617-75), who had the advantage of some schooling as a pupil of Galileo, was also an amateur scientist. In due course, he was persuaded by another of Galileo's pupils, Vincenzo Viviani (1622-1703), and by Giovanni Alfonso Borelli (1608-79), to stand as founder, patron, and first president of the Accademia del Cimento. This "academy of experiment" did not grow in size beyond the original nine members recruited by Borelli, Viviani, and Carlo Rinaldini (1615-98),[68] and it survived as an organization for only ten years. It was unique, however, in two respects: it was the first formally organized society devoted primarily to the promotion of *experimental* physical science and the first to demonstrate the effectiveness of sponsored team attack on major problems in science.

The studies undertaken by the Accademia were truly distinguished for their time and included such topics as the Toricellian barometer and its associated "Toricellian vacuum," the freezing of liquids, thermometry and the transmission of heat, the propagation of sound and light, and the phenomena of magnetic and electrical attraction. Anonymity within the group was a general rule and it is necessary to rely on the manuscript journals and correspondence underlying the "final report"[69-71] of 1667, prepared by Magalotti as secretary, for information as to which of the academicians were involved in which experiments. In at least a few instances, the academy appears to have taken over and "adopted" investigations that had already been carried out, or at least started, even before the academy was formally established in 1657.

One of the topics of this category of adopted work was a redetermination of the speed of sound. The grand duke Ferdinand II had "made certain experiments with gunfire at Petraia near Florence," and had called on his senior scientific protégé Borelli and on Viviani to help in measuring more accurately the distance involved in these tests. Their methods for doing this, incidentally, must have been relatively crude even for that period, inasmuch as the uncertainties in dispute were substantial. A further series of tests with gunfire was then planned in collaboration with Leopold de Medici and carried out by Borelli and Viviani during the night of 10 October 1656. The primary aim of these trials seems to have been to prove again that Mersenne and Gassendi were right in asserting the like speed of sounds weak and strong, with or against the wind, and over the first and last half of any given course. That these one wrong and two right conclusions were impartially "proved" gives support to the idea that it was not their quantitative results that were important, any more than Mersenne's had been; what was important was the fact that quantitative experiments were undertaken at all.

The tests were carried out with Borelli stationed at one end of "the new avenue of Bosco di San Moro dalle Mullina" and with Viviani at the midpoint, and each had a pendulum "which made the one and the other equal vibrations." To Viviani's relief, since he had predicted such a result, exactly twice as many vibrations (between the flash and the arrival of the sound) were counted by Borelli's party at the distant station as by Viviani. As for the speed itself, this was given only in terms of an ambiguous statement that "in 15 1/2 vibrations, the sound traversed 1 1/5 miles, i.e., 3600 *braccia*";[70a] but nothing was said explicitly about the length or period of the pendulum. This omission is the more surprising because Viviani must already have been pendulum-conscious as a result of his association with Galileo and Galileo's son Vincenzio Galilei (1606–49), for whom the *Saggi* made a spurious claim that he had successfully applied the pendulum to the movement of the clock just before his death in 1649. Moreover, it was probably Viviani who devised the ingenious bifilar pendulum suspension[69a] described in the *Saggi*, which made the pendulum more useful as a timekeeper by constraining it to oscillate in a single vertical plane. If the pendulum used in these 1656 trials was the same half-second "dondolo" mentioned explicitly in the description of the suggestively similar experiment reported later in the *Saggi*, then their 1656 distance measurements must still have contained gross errors.

As suggested above, neither the dates nor the experimenters are identified in the *Saggi*, but it seems reasonable to assume that Viviani and Borelli are to be credited with the academy's measurement of the speed of sound, whether the *Saggi* account refers to a new experiment or to a reinterpretation of the 1656 trials. Waller's free English translation of the *Saggi*, with its quaint mixture of italics and surprising capitalization, manages somehow to capture the mood as well as the substance of these experiments:

At the distance of one of our *Miles* exactly measured, which are about 3000 of our *Braccia*, or 5925 Foot, we fired several Pieces, that is Six *Harquebusses*, and as many *Chambers;* at each whereof from the *Flash* to the arrival of the *Report*, we counted Ten whole *Vibrations* of the *Pendulum*, each for which was half a *second*. Repeating the *Experiment* at half a Miles distance, that is, at the *mid way*, we observed it to be exactly in half the time, always counting at each Report, about five *Vibrations*, wherefore we rested satisfied of the certainty of the *equability*.

The *Consequences* which we pretend will follow from this *equability*, amongst the rest are, That by the Flash and Sound of divers *shot*, we might obtain an exact Measure of the Distance of places; particularly at *Sea*, of *Ships*, *Rocks*, and *Isles*, where we cannot come to take several *bearings*, as is requisite in using the common *Instruments*: we may also by a single stroak made upon *Wood*, *Stone*, or *Metal*, or any other sounding Body, judge how far off he is that gives the *blow*; . . .

If we would likewise know the *Distance* of Places, which because of the *Roundness* of the Earth, or interposition of *Hills*, we cannot have a sight of, yet with ease we may obtain it, and that by *Two Discharges*, answering each other; so that to our *firing* at one place, they

must return *another* at the other place; and taking the middle time
between our *discharge*, and the arrival of their *Answer*, the half of the
Sounds Journey will be found, that is, the whole Distance of the
Places sought.

By the same way of *Sounds*, the *Maps* of particular Places may be
adjusted, and truly laid down in *plans*; taking first the *Angles* of Position of the *Cities*, *Castles*, and *Villages*, to place them in their due
scituation; with several the like curious *Inventions* very useful, nor to
be disesteemed.

Then to gain the unknown Distances of each, we may make use of
time for a Scale, the *sound* travelling with us the known space of a
Mile in Five *seconds*.[69b]

The conversion of "one of our miles" of 3,000 *braccia* into
5,925 feet was Waller's interpolation. If it is assumed, perhaps
gratuitously, that these were the same as modern English feet
of 30.48 cm. the corresponding speed of sound would be
361 m/s. This would seem to indicate that their difficulties
with distance measurement may indeed have been cleared up
finally since this result is only 5 percent too high.

The proposal that sound-travel time be used for a distance
scale was not new. Mersenne had already broached this idea
and had embroidered and extended the notion further by disclosing the first conception of the principle of echo ranging.[54s] On the other hand, the suggestion that a second answering discharge be used as a "transponder" or "echo re-
peater" was both a novel and a sound variation on the echo-
ranging theme, the value of which is not lessened by the fact
that it had to lie dormant for nearly three centuries until it
could be implemented by the radar and sonar techniques
developed during the Second World War.

It may well be that the most important experiment undertaken by the Florentine academy was not concerned with
physics at all but with its own attempt to survive, for the fact
could hardly be disguised that its guiding principles were in
direct conflict with the church-endorsed deductive science of
the classical quadrivium (arithmetic, music, geometry, and
astronomy). Various reasons were subsequently advanced to

explain the premature dissolving of the academy: did the members fall out among themselves, or was it true, as Magalotti charged, that dissolution of the academy was the price paid by its patron for a cardinal's hat?[70b] It is at least factual that Leopold *did* become a cardinal in 1667 and that the short-lived Accademia del Cimento expired in the same year, after Leopold had resigned and withdrawn his patronage.

The Renaissance spirit of intellectual protest that flared up briefly in the Italian academy survived more stoutly in the manifestations of collective scientific action appearing elsewhere, in England and on the Continent, during the seventeenth century. Both the Royal Society of London and the Parisian Académie des Sciences were formally incorporated during the brief period in which the Accademia del Cimento still flourished; but like the latter, each had really come into being a decade or two earlier as a loosely organized gathering of men interested in experimental science and motivated by a common desire to share their results and to hear about the work of others.

Of the origins of the Royal Society the mathematician John Wallis (1616-1703) recalled in 1696 that "About the year 1645 . . . divers worthy persons, inquisitive into natural philosophy and other parts of human learning, . . . did by agreement, diverse of us, meet weekly in London on a certain day, to treat and discourse of such affairs."[72] Paralleling these London meetings, there were also weekly meetings at Oxford of what Robert Boyle (1627-91) called in letters of 1646-47, the "invisible college," and this group associated itself in 1648 as the "Philosophical Societie of Oxford" which survived until 1690 (cf. n. 60). As circumstances offered, there was a free interchange of personnel between the Oxford and London groups, and many of the former were present at the meeting of 28 November 1660 at Gresham College in London, when "it was proposed . . . to improve this meeting to a more regular way of debating things, and according to the manner in other countrys where there were voluntary associations of men in academies for the advancement of various parts of learning."[73] Dr. John Wilkins (1614–72) was named as temporary chairman, lists were drawn up of persons "judged willing and fit to join with them in their design," informal rules of procedure were drafted, and Sir Robert Moray [or Murray] (1600–73) sought the favor of King Charles II (1630–85) with such good effect that a Royal Charter dated 15 July 1662 duly established the Royal Society of London for the Improving of Natural Knowledge.

The French counterpart of Boyle's "invisible college" was the series of séances convened more or less weekly at the Hôtel des Minimes in Father Mersenne's quarters and attended by such men as Pierre de Fermat (1601–65), Gilles Personne de Roberval (1602–75), Pierre Gassendi, Etienne Pascal (1588–1651), Blaise Pascal (1632–62), and such occasional visitors from abroad as Thomas Hobbes (1588–1679), Giovanni Domenico Cassini (1625–1712), and Christiaan Huygens (1629–95). The opportunity these weekly meetings provided for the exchange of information was made richer by the knack Mersenne had for keeping up a sympathetic exchange of letters with leading natural scientists almost everywhere on the Continent. The seventeenth-century equivalent of the modern practice of announcing new results by a "letter to the Editor" of a scientific journal was to write a personal letter to Father Mersenne. It was in this way, for example, that the news of Toricelli's research on the vacuum (1644) first reached such French scholars as Pascal.

After Mersenne's death, these meetings were held for a few years at the house of Henri Louis Habert de Montmor (ca. 1600–79), to whom Mersenne had dedicated both the 1636 and 1648 editions of his *Harmonicorum libri*, and later at the home of Melchisédec Thévenot (1620–92). The statesman and organizer, Jean Baptiste Colbert (1619–83), learned of these meetings from Claude Perrault (1613–88) and interceded to gain from Louis XIV (1638–1715) a royal charter under which the Académie des Sciences held its first formal meeting on 22 December 1666. Unlike the Royal Society, the Paris academy enjoyed a substantial measure of "government support," and its members were granted both pensions and living quarters. Perhaps as a consequence of this dependence on the public treasury, the Académie des Sciences suffered several reorganizations and deviations of design before it was finally reconstituted as a branch of the Institut National in 1816.

Hardly less important than the forums provided by these academies, and of more lasting significance, were the scientific journals they sponsored. The *Philosophical Transactions* of the Royal Society was launched on 6 March 1664/5, under a license from the council of the society but at the private risk of the second secretary, Henry Oldenburg (1626-78). Some intermittency of publication followed Oldenburg's death, one gap of two years being filled by the issuance of Hooke's *Philosophical Collections.* Regularity was reestablished in 1691, and the *"Phil. Trans."* has been published without interruption throughout the almost three centuries since that date, thereby establishing an unparalleled record of publication service to science.

The French Academy, on the other hand, did not intend at the outset to publish its own journal, perhaps because Denis de Sallo (1626-69), a French parliamentarian and satellite of Colbert, had brought out privately the first number of an independent *Journal des Scavans* on 5 January 1665, just a few months before the appearance of the first issue of the *Philosophical Transactions*, and had elected to report rather fully on the activities of the academy. This loose and unofficial connection between the *Journal des Scavans* and the academy was tightened and eventually formalized after Sallo's license had been withdrawn four months later and the direction of the *Journal* had been taken over by a member of the academy, l'Abbé Jean Gallois (1632-1707). The title of this publication was changed from time to time as the Académie des Sciences went through its successive stages of reorganization, but it continues to share with the *Philosophical Transactions* the distinction of being the first of the "modern" scientific periodical journals.

The subsequent growth of academies and learned societies in almost every quarter where scientific learning was respected is an outstanding feature of the evolution of science. It is defensible, in fact, to claim that a history of the scientific societies and their journals after 1665 would be an adequate, if not inclusive, history of post-Galilean science itself. Only two of these later societies will be mentioned explicitly here; the rest will be allowed to make their appearances in the footnote references. One was the Collegium Curiosum sive Experimentale established in 1672 by the mathematician and physicist Johann Christoph Sturm (1635-1703) at the progressive German University of Altdorf (merged with Erlangen in 1809). This society was ostensibly fashioned on the pattern of the Accademia del Cimento, but its chief distinction stems from the fact that it was the first to be devoted to fostering scientific research in a university. It did not survive many years after its founder died, but it was the prototype of all the modern professional societies devoted to special branches of science.

The other pioneer learned society was the Berlin Academy which, after the Royal Society and the French Academy, was the most permanent of those established during the seventeenth and eighteenth centuries. Unlike its predecessors that had "grass roots" origins, the Berlin Academy was the almost single-handed creation of Gottfried Wilhelm von Leibnitz (1646-1716), who showed a passion for organizing scientific societies that almost matched his mathematical genius (of which more later). Leibnitz did not remain attached for very long to any of the organizations he was instrumental in starting, although he would probably have found most satisfaction in the Academy of St. Petersburg, whose founding in 1724 can be traced to earlier suggestions he had made to Peter the Great (1672-1725).[74]

The Swiftness and the Medium of Sounds

The measurement of the speed of sound undertaken (or taken over) by the Florentine Academy soon found a counterpart in similar activity sponsored by the Royal Society and the Paris Academy. Newton's theoretical derivation of the speed (in the *Principia*, about which more later) was the first example of deductive inference of a basic physical constant, but its publication only rekindled interest in further experiments, since the speed predicted by Newton's theory was in serious disagreement with all the measurements that had been made up to that time. With some hope of resolving this dilemma, the Royal Society requested John Flamsteed (1646–1719), the Astronomer Royal, and Edmund Halley (1656–1742), who succeeded Flamsteed in that post in 1720, to undertake a careful redetermination of the speed. There is reason to suspect that this may not have been a happy collaboration, as Flamsteed's regard for Newton deteriorated to enmity before the turn of the century, whereas Halley had already objectively demonstrated his warm sympathy for Newton by securing with his own funds the first publication of the *Principia*. Nevertheless, the measurements must have been carried out, since Derham referred to results that he ascribed to them in his own report to the Royal Society on the same subject in 1708; but no joint report by Flamsteed and Halley seems to have been made.

As one of its early activities, the Académie des Sciences also undertook a measurement of the speed of sound that was carried out on 23 June 1677 by G. D. Cassini, Jean Picard (1620–82), and Olof Römer (1644–1710), the latter having gained the distinction two years earlier of being the first to observe the finite speed of light by measuring the difference in the times at which Jupiter's moons are eclipsed when the earth is at different points in its orbit. This team of astronomers was then primarily engaged in erecting and putting into operation the Paris Observatory, with the result that their measurement of sound speed must have been undertaken rather casually, if one can judge by the fact that Du Hamel devoted a mere five lines to a description of the experiment in his history of the first thirty-four years of the Royal Academy of Science.[75] Their results were very creditable, nevertheless, seven seconds of a minute being counted while the sound of a *bombard* covered the "*1280 hexapedas* between the Observatory and a place called Glarea [later the *Quai de Grève*, now part of the *Quai de l'Hôtel de Ville*]," which corresponds to a sound speed of 356 m/s.

Since the Newtonian dilemma concerning the speed was still unresolved half a century later, the Paris Academy in 1738 sponsored another determination that was to stand up as the most reliable measurement of sound speed made during the eighteenth century. Giovanni Domenico Maraldi (1709–88) and l'Abbé Nicholas Louis de la Caille (1713–62) were members of the 1738 measurement team, which was led by (César François) Cassini de Thury (1714–62). The latter was a grandson of the G. D. Cassini who had headed the 1677 team and was later to become the third of the four generations of Cassinis to direct the Paris Observatory. The 1738 experiment was conducted with much circumspection: cannon were fired alternately at the two ends of an 18-mile base line in order to allow the effects of prevailing winds to be eliminated, and care was taken to record the temperature at each observation post even though no effect of temperature on sound speed was then known to exist.[76] The availability of these data made it possible to reassess the measurements later and to deduce from them a value of 332 m/s for the speed of sound in air at 0° C., in excellent agreement with modern determinations.

The fact that temperature does influence the speed of sound was not established until about 1740, when two experimenters working independently each found the opportunity to compare measurements made at different ambient temperatures. Count Giovanni Lodovico Bianconi (1717–81) made measurements at Bologna in 1740 over the same base line, first in winter and then in summer, and concluded that the speed of sound in air increases as the temperature rises.[77] This general result was soon confirmed by the somewhat less reliable observations of Charles Marie de la Condamine (1701–73), who compared his measurements made in 1740 at high and cool Quito (which then belonged to Peru) with those he made later in 1744 at much warmer Cayenne (in French Guiana).[78]

This experimental evidence of the influence of temperature on the speed of sound did not attract much attention when it first appeared, inasmuch as the failure of Newton's theory to account for the speed of propagation at any temperature was still an unsolved puzzle. It turned out, of course, although not for another seventy years, that the key to the Newtonian dilemma was the different elasticity manifested by air under adiabatic and under isothermal compression; and when this difference came to be understood, there came with it almost automatically an understanding of the effect of temperature on the speed of sound. As a result, making allowance for differences in temperature when analyzing data on sound speed became simply a matter of applying a simple correction rather than a subject for independent experimental investigation. This was far from the case, however, at the turn of the eighteenth century, and most of the repetitions of the sound speed experiment were motivated less by the urge to achieve greater accuracy than by a desire to learn more about the fundamental nature of sound propagation itself.

One can wonder with some humility whether any of the twentieth-century questions concerning the atomic nucleus will ever sound as naïve as the series of acoustical questions to which William Derham (1657–1735), the rector of a small church near London, addressed himself in 1704.[79] Was sound propagated vertically at the same speed it traveled horizontally? Did it make any difference whether the cannon was pointed toward or away from the observer (assuming him to be out of range of the shot)? Did it matter whether the barometer was rising or falling? What was the effect of the wind? And how did temperature and humidity affect transmission? Derham found correct answers to most of these questions, although he allowed the equability of English weather to trap him into concluding that the state of the atmosphere, as well as the time of day, was without influence on the speed. Incidentally, Derham also joined the growing ranks of those who proposed some form of the scheme for using sound travel-time for distance measurement when he suggested that one could tell the remoteness of a distant thunderstorm by observing the time interval between a flash of lightning and the sound of its thunder.

The bell-in-a-vacuum experiment has served in elementary physics classes for many years as a dramatic demonstration that sound is interdicted when the surrounding air is withdrawn from a vibrating body. The confused interpretation of this experiment, however, when it made its first appearance in 1615 and on the occasions of many of its repetitions during the following century, epitomized the prevailing level of insight regarding the medium of propagation for sound waves. Gianfrancesco Sagredo was the first to describe such an experiment, which he carried out by suspending a "hawking bell" in a glass vessel that was sealed off while it was still very hot. This procedure could not have reduced the residual air pressure to much less than a third of an atmosphere, so it can be doubted whether his observations were truly conclusive. Nevertheless, the effect was sufficiently striking to support him in a curious inversion of the logic: "I persuaded all present," Sagredo wrote in a letter to Galileo, "that there was in it [the vessel] very little air," since the bell "when moved, made no sound."[80]

It did not take very long, as time marched in those days, for Sagredo's demonstration to "catch on." Isaac Beeckman (1588–1637), in his autographic *Journal*, mentioned in passing that he had seen a similar experiment performed by an Italian who was touring the Low Countries in December

1629.[81] Confusion regarding both motive and interpretation still surrounded the experiment, however, even in the much fuller treatment it received a decade later at the hands of Gasparo Berti (sometimes Alberti; d. 1643) and Athanasius Kircher (1602–80), the latter a professor of mathematics in the Jesuit College of Rome (now the Pontificia Universitas Gregoriana).

Kircher's elaborate account of these experiments is set forth in his long treatise *Musurgia Universalis*, published at Rome in 1650. After beginning his *Digressio*, subtitled "Whether a sound can be made in a vacuum," with a description of the mercury barometer Evangelista Torricelli (1608–47) had used to demonstrate the weight of the atmosphere,[82] Kircher bitterly castigated those who "argue that the space left in the upper part of the tube is truly and properly a vacuum." Then, after some fanciful speculation about what might be there instead of a vacuum, he pretended to clinch the argument by asserting "that it is quite impossible for there to be a vacuum there, since a very distinct sound is perceived in it: a fact that I found out by trial several years ago, along with Gaspar Berthius [= Berti],[83] a most talented Mathematician here in Rome."[84] This was, of course, a blatantly improper way to argue *whether* sound can be made in a vacuum, and Kircher did little to mend the illogic in what followed.

As for the "trial," which must have been made between 1640 and 1643, Kircher claimed that "At my suggestion, he [Berti] inserted on the side of the phial [in the 'vacuum' space at the top of a long lead tube] a bell together with a hammer, cleverly arranged so that the iron hammer, when attracted and raised by a magnet from outside and then freed, would by its own weight produce the sound of the bell." When the long tube had been first filled with water and then inverted:

the water descending within the lead tube did not discharge itself entirely . . . whence many spectators of this so wonderful experiment tried to infer that the space abandoned by the water was necessarily a vacuum, since no other body could have been substituted in the same place. But in order to show by actual aural experience the falsity of their opinion, we seized a magnet, applied it to the glass phial outside in the vicinity of the iron hammer and raised the attracted hammer, but when the magnet was withdrawn the hammer by its own weight produced a very clear sound from the bell. Whence certain more obstinate-minded persons who were present at once inferred that sound could be made in a vacuum; but the more sensible deduced at once from this very clear experiment that it was quite impossible for there to be a vacuum in a place where such manifest signs of air were displayed in the sound.

Kircher failed to strengthen the argument in his extension of the foregoing exposition, but he did reveal his mixed reliance on the spellbinder's tricks of repetition, distraction, and counterattack:

> The non-existence of a vacuum is most clear from what has been said, but if anyone should ask me by what method or by what hidden ways the air substitutes itself in the place of the departing water, I should satisfy him too, if he would first explain to me by what method the magnet penetrates glass and other hard bodies, and light penetrates the most solid crystal; for the paths of struggling nature are so secret, so hidden, that they are quite impossible of comprehension by the human mind. Nature is so ingenious in time of necessity, that she seems to elude all scrutiny of the human intellect; at any rate the experiment makes it certain that air got in there, but how and where is not clear.

Finally, with a syllogistic flurry reminiscent of the reasoning by which Aristotle "proved" the nonexistence of a vacuum by the nonappearance of like and infinite rates of fall for light and heavy objects, Kircher concluded "that even if a vacuum were possible in the nature of things, nevertheless sound could not occur in it. For since sound is an affection of the air—in fact air is the material cause of sound—, when that is lacking, sound also must be lacking; and on the other hand, we have clearly shown from the proposed experiment that a vacuum cannot be admitted in the nature of things."[84]

Within a few more years the bell-in-a-vacuum experiment

had been repeated in three widely separated places with as many widely differing results. The first after Kircher and Berti was probably the "Experiment of Sounds in Vacuo" reported in the *Saggi* of the Accademia del Cimento, although there is the usual ten-year uncertainty about dating the work of the Florentine academy. Waller's translation describes this experiment as follows:

> Having hung a small Bell by the thread, . . . and making the vacuum, we began to shake the Ball [i.e., the glass "receiver" corresponding to the "phial" at the top of Berti's tube] forcibly, and the Bell gave the same Tone as if the Ball had been full of common *Air*; or if there was any Difference it was too little to be perceived; indeed, in this *Experiment* the sonorous *Instrument* (tho the thing is impracticable) ought to have no communication with the Vessel, otherwise we cannot certainly affirm, whether the Sound proceeds from the Rarified air, and *Effluvia* of the *Mercury* in *Vacuo*, or from the Vibration communicated by means of the Thread from the percussion of the Metal to the Glass, and so to the External *air* encompassing it.[69C]

There are two notable things about this account besides the wrong conclusion it reaches: the reference to a Torricellian vacuum containing the "effluvia of the mercury," which the academicians may have understood better than Waller did, since a more literal translation of the Italian would be "vapor evaporated into the vacuum from the mercury"; and the clear understanding of the requirement that there be no secondary channels of communication between the "sonorous instrument" and the "vessel." To recognize the latter requirement was wise, but to meet it was [and still is!] difficult; and one can infer from the results that their technique of mechanical isolation was less than adequate.

Before the academicians drew a summary conclusion, another, and rather bizarre, variation of the experiment was carried out. A small organ pipe was connected to a bellows and both were mounted inside a cylindrical copper "vacuum chamber." The bellows could be operated from outside by a long handle projecting through a flexible gasket, while the "vacuum" was produced by a primitive form of hand-operated piston pump. One can infer that this pump was as relatively inadequate as the mechanical isolation had been in the bell experiment, since the pipe continued to sound when the pump was operated, although the nature of the sound did change. "Wherefore," the academicians were forced to conclude "(some did say jestingly, either that) the *air* has nothing to do in the Production of *Sounds*, or is able to do it alike in *any state*" (see n. 69c).

The Italian academicians might have reached a different conclusion if they had used one of the series of progressively improved air pumps devised from about 1654 on by Otto von Guericke (1602–86), the burgomeister of Magdeburg. In his version of the experiment, Guericke suspended in the recipient (receiver)

> a mechanical timepiece with a high-pitched ring . . . arranged in advance so that by the striking of the hammer on the bell it should emit a sound separated by fixed intervals over a half-hour. When this had been done, and the recipient stoppered, I proceeded to extract the air, and I noted that when the air was partially pumped out the sound came out weaker, and when the air was completely exhausted it did not even reach my hearing. However, if I moved my ear close to the glass, a dull noise from the bell, rising from the stroke of the hammer, stimulated my ear, just as, if one should clasp the whole of a bell of this sort in his hand and strike it with some hammer, he perceives a duller sound or noise produced by the contact, but by no means a ringing. . . . Hence we recognize that sonorous objects, such as bells, cymbals, glasses, and strings of musical instruments and other things of that sort, produce their ringing by benefit of air, that is by the vibration or trembling with which they beat the air; on the other hand, the noise or din which is aroused by mere friction or rubbing together of things is aroused not through the medium of air, but by the Sounding Virtue itself.[85a]

Guericke seems, on the whole, to have been more impressed by the "noise and din" that was produced in a vacuum than by the ringing that was not, and the "Sounding Virtue" to which he referred led him into some strange pathways of speculation elsewhere in his text.[85b]

When Schott's description of Guericke's pump reached

England, it came to the attention of Robert Boyle (1627–91), and he set about at once, with the help of his assistant Robert Hooke (1635–1703), to modify and improve it. In his first experiment with sounds in a vacuum, Boyle chose a watch as the sound source, and this

> was suspended in the cavity of the Vessel onely by a Packthred, as the unlikliest thing to convey a sound to the top of the Receiver: . . . The Pump after this being imployd, it seemed that from time to time the sound grew fainter and fainter; so that when the Receiver was empty'd . . . neither we, nor some strangers that chanc'd to be then in the room, could, by applying our Ears to the very sides, hear any noise from within; though we could easily perceive that by the moving of the hand which mark'd the second minutes, and by that of the ballance, that the watch neither stood still, nor remarkably varied from its wonted motion. And to satisfy ourselves farther that it was indeed the absence of the Air about the Watch that hinder'd us from hearing it, we let in the external Air at the Stop-cock, and then . . . we could plainly hear the noise made by the ballance, though we held our Ears sometimes at two Foot distance from the outside of the Receiver. . . . Which seems to prove, that whether or no the Air be the onely, it is at least, the principal medium of Sounds.[86]

This deduction was guarded, but it was less equivocal than Guericke's in spite of the similarity of the two experiments and the fact that they must have been performed almost concurrently. Their conclusions may have differed because Boyle's "watch" was a more feeble sound source than Guericke's "timepiece with a high-pitched ring," and therefore more easily stilled; but it is equally likely that Hooke's improvements had made Boyle's pump more effective than Guericke's, from which it had been copied, and that there were real differences in the experimental phenomena they observed as well as in their conclusions.

There was also a sharp contrast between Guericke's fanciful allusions to the "sounding virtue" and Boyle's lucid description of the mechanism of sound transmission through the walls of the receiver before the air was pumped out. This he explained in terms of sound waves beating on the walls of the receiver and setting them in vibration, with consequent re-radiation of the sound into the external air beyond, an explanation that is still valid as a characterization of the dominant mode of sound transmission through walls and building partitions. Boyle tried another experiment in which a small bell was supported by a bent stick pressing against the sides of the receiver. This was heard whether the air was exhausted from the receiver or not and, while Boyle thought this was a plausible result of the direct mechanical connection, it seems to have tempered the firmness with which he declared the uniqueness of air as the "medium of Sounds."

Francis Hauksbee (d. 1713) took up the question of sound in a vacuum about half a century later and devised several modifications of the classical experiment. He is often credited with giving the first definitive proof that a material medium is needed for the transmission of sound waves, although his experiments were hardly more conclusive than Boyle's in this respect. Nevertheless, his experiments did attract more attention, perhaps because they were performed before the Royal Society and were reported in the *Philosophical Transactions.*[87]

Hauksbee's first trials in 1705 included a repetition of the conventional bell-in-a-vacuum experiment and the execution of one of the tests that Boyle had proposed but did not carry out, in which the air pressure in the receiver containing the bell was increased rather than decreased. In each case the bell was rung by shaking the apparatus to make the clapper strike. Hauksbee pointed out explicitly that "it was very observable that the Interposition of the Glass [i.e., the walls of the receiver] betwixt the Bell and the Ear, was a great obstruction to its Sound, notwithstanding it was audible at some great distance." Then, as usual, on "gradually withdrawing the Air, and making several Stops to shake the Bell at different degrees of Rarefaction, the Diminution of the Sound at every Stop was very distinguishable. Till at last, when the

Receiver was well exhausted of Air, the remains of Sound was then so little, that the best Ears could but just distinguish it: It appearing to them like a small shrill Sound as at a great remoteness."[87a]

In the converse experiment, "some Gentlemen to observe the Sound" were stationed at the other end of a long room. "Before any Air was intruded, the Bell upon shaking was heard at that distance, tho not without diligent attention. Upon the Intrusion of one Atmosphere (begging leave to call it so) the Bell being shaken as before, the sound was very sensibly augmented."[87b]

Hauksbee's pump was not equal to the task of raising the excess pressure much beyond two atmospheres, but this did not deter him from pressing his efforts to find a quantitative measure of the effect of condensing the air. When he tackled this problem in a follow-up experiment, Hauksbee joined the fraternity of hardy and sleepless experimenters (of which Viviani and Borelli had become charter members with their 1656 speed-of-sound tests) who continue to find that the still small hours of the night offer the most favorable acoustical environment for outdoor experiments. Hauksbee's testimony ran as follows:

> About 5 in the morning, I repeated this Experiment in an open Field, known by the name of the *White Conduit*, situate on the West of *Islington*, with much the like success as the former. Upon shaking the Bell before any Air was intruded, it was but just audible at 30 yards distance. Upon the Injection of one Atmosphere (begging leave as before to call it so) it became then as audible at 60 yards, as it was before at 30. A second being intruded, the Bell upon shaking might then be heard at 90 yards distance (see n. 87b).

It often happens, as in this case, that the most interesting feature of a new experiment is the method used rather than the results obtained. Mersenne had relied on a similar line of reasoning in estimating the distance at which artillery could be heard, but Hauksbee deserves much credit for so explicit an example of using a loudness judgment in conjunction with the inverse square law of energy transmission as a measure of source strength.

The record does not show whether Hauksbee's colleagues criticized his results or whether his misgivings were self-generated, but he did raise the question "whether the Sonorous Body in such a *Medium* [vacuum] might not suffer, or undergo such a Change in its Parts, as to be render'd uncapable of being put into such a Motion as is requisite for the Action or Production of Sound." In order "to set the Matter of Fact in a true Light," Hauksbee mounted the bell "as large as well it could contain" in a receiver and then covered this with a still larger receiver arranged so that the air could be exhausted from the space between the two. "Now here I was sure," Hauksbee declared, that "when the Clapper should be made to strike the Bell, there would be actual Sound produced in the inward Receiver, the Air in which was of the same density with common Air." Then after pumping out the space between the receivers, and "all being ready for Trial, the Clapper was made to strike the Bell; but I found that there was no transmission of it thro the *Vacuum*, tho' I was sure there was actual Sound produced in the Receiver. This plainly shews, and seems positively to confirm, That Air is the only *Medium* for the Propagation of Sound."[87c]

Hauksbee was obviously gratified by the outcome of this modification of the test, which he regarded as crucial; and the firmness of his conclusion bolstered the opinion that the test was definitive. Yet nothing really significant was added to the earlier experiments by the modification, which amounted to no more than replacing the vibrations of the bell by the vibrations of the wall of an air-filled receiver containing the bell. On the other hand, Hauksbee's misgivings about the validity of the previous experiments were well founded, although his qualms about the effect of a vacuum on the sounding source might better have been turned around to bear on the question of whether there could be any effective interaction between the vibrating source and a highly rarefied medium. Actually to "set the Matter of Fact in a true Light," it must be said that an erroneous interpretation has adhered to these experiments with stubborn tenacity from the time of their first performance in the seventeenth century down to their present and continued use as lecture demonstrations in elementary physics classes. As Lindsay recently reminded physics teachers, these experiments do *not* prove the inability of a rarefied medium to *transmit* acoustic energy, but only the extreme difficulty of *imparting* any appreciable amount of vibratory energy to such a medium.[88]

Acoustical Technology and the Horn

Kircher's voluminous *Musurgia* (see n. 84) of 1650 contained many items of acoustical interest in addition to the account of his bell-in-a-vacuum experiments. Several of his proposals were speculative and fanciful, but they attest to Kircher's fertile imagination if not to his mechanical and acoustical sophistication. He had much to say about the production of "voice," the anatomy of the vocal organs and the ear, and the generation of musical sounds, but what he said added little to the state of knowledge relating to these topics. On the other hand, his description of ancient musical instruments constituted a useful contribution and some of his illustrative drawings have been widely copied.

Kircher was at his imaginative best in the sections of the *Musurgia* devoted to automatic players for bells, organs, and strings.[89a] The bells and organ pipes were to be controlled by levers actuated by teeth projecting from a revolving drum (*cylindrum phonotactica*). A chart representing the developed surface of the drum was proposed as a vehicle for transcribing a musical score into the corresponding positions of the raised teeth, in much the same way that this art must still be practiced in the construction of "music boxes." His "new and perhaps unheard-of *clavicembalo* which will by its sound produce a symphony of viols"[89b] was equally ingenious.

Several strings were to be stretched over arched bridges so that each would be near, but not quite in contact with, a rotating wheel. Then, as one or another key was pressed, the corresponding string was to be "bowed" by being drawn into contact with the rotating wheel (see n. 89a). There is no evidence, unfortunately, to indicate that any of these automatic players was ever constructed. Another proposed example of automation was the process that Kircher called "Mechanical Musurgy," a kind of slide-rule "method invented by us whereby anyone, however unmusical, can compose melodies by the various use of tune-making instruments" (n. 89b).

Kircher also concerned himself with the Aeolian harp, a model of which was displayed in his museum at Rome. Some writers have credited Kircher with the invention of this instrument, but he did not claim it; nor, it might be added, did he seem to understand it. At least, his explanation of how the wind impinging on only a portion of the string could cause it to emit tones at other than its fundamental or harmonic frequencies failed to qualify him as a theorist.[89c]

Neither can Kircher be ranked very high as an experimentalist, if one can judge this by his unsuccessful attempts to repeat two of Mersenne's classic experiments. He quoted one of Mersenne's examples of the model string experiment for frequency determination (without acknowledging that they were Mersenne's numbers), but he did not seem to grasp fully Mersenne's principle of modeling. In Kircher's version of the experiment, he seems to have tried to count the vibrations of the short segment of the long string, and so it is not surprising that he found the motion of the chord to be "so rapid, confusing, and indistinct that it dashed all my hopes of computing the vibrations. Yet the vibrations can surely, as I have said, be known by hypothesis or supposition."[89d] Apparently the motion of the chord was not the only thing that was confused! Kircher had no better luck when he tried to repeat Mersenne's echo experiments for determining the speed of sound, and he could only conclude that sound had many speeds, being faster when loud or at midnight and in the evening, and slowest in the early morning.[89e]

The work of Kircher that commands the most respect is that dealing with the acoustical horn and certain limited aspects of architectural acoustics. The material on these topics contained in his *Musurgia Universalis* of 1650 was extracted and republished in 1673 under the title *Phonurgia Nova*,[90] the expansion of the earlier text occurring chiefly in the parts dealing with the horn. Several of the drawings and some of the text taken from the 1650 work were rather carelessly copied, from which it can be inferred that the *Phonurgia Nova* may have been published in haste, perhaps in order to press Kircher's claim to prior discovery of the loud-speaking trumpet that Morland had disclosed in an English publication of 1672 (see below).

Kircher's work on architectural acoustics was primarily concerned with the geometrical design of rooms to provide sound-focusing effects. Ray diagrams had already been used in the analysis of focusing effects by Mersenne in the field of acoustics and by many others earlier in the field of optics, but Kircher was the first to extend this concept to architectural design.[89f, 90a] He also discussed at some length the rationale of using vases of assorted sizes in theaters as Vitruvius had recommended, but neither his experiments nor his discussion proved that he understood this problem any better than Vitruvius had.[89g, 90b]

Kircher's most successful experiment, and the one on which his claim to be the inventor of the loud-speaking trumpet rests, involved a conical horn 22 palms long (about 16 feet if one takes the *palma* to be about 8.7 inches). This was made of iron plates and extended from a small (2-inch) opening in one wall of his workroom to a larger (2-foot) aperture in an outside wall facing the garden courtyard of his quarters in Rome.[90c] He could use this tapered conduit either as a speaking-trumpet or megaphone through which he could talk to the gatekeeper without leaving his quarters, or as an ear-trumpet by means of which he could eavesdrop on conversations taking place in the courtyard—all to the amazement of various auditors who were invited to share the experience at one or the other terminal of this acoustical transmission link.

Kircher discussed at some length the possible effects of giving various shapes to the horn, but his theorizing was unsound and did little to advance this art. Needless to add, this fact did not divert him from proceeding to a variety of bizarre elaborations of his through-the-wall trumpet for such uses as making statues appear to speak, for "broadcasting" the music of players in an inside room to dancers in a courtyard, or for eavesdropping in the classic manner of the tyrant Dionysius the Elder (ca. 430–367 B.C.).[89h, 90d] Perhaps stimulated by Morland's activity, Kircher made and tested (ca. 1672) a portable (?!) loud-speaking trumpet nearly 10 feet long and 3 feet in diameter whose sound-carrying power was demonstrated by the use of it to summon some 2,200 people from as far as four miles away to a special church service![90e]

The English work on a speaking-trumpet which triggered off Kircher's claim to priority was described by Sir Samuel Morland (1625–95) in a brochure entitled *Tuba Stenotoro-Phonica . . . ; Invented and variously Experimented in the Year 1670.*[91] Morland's first instrument was made of glass, 32 inches long and 11 inches in diameter at the mouth, and from this he progressed to a copper 21-footer with a 2-foot mouth. Such a long horn should indeed have delivered impressive performance, although Morland may have stretched the facts a bit in his enthusiastic

Account of the Manifold Uses of this Tuba Stentoro-Phonica . . . At Sea, In a Storm, or in a dark night, when two Ships dare not come so near one to the other as to be heard by any mans ordinary Voice; . . . At Land. In case a Town or City be Besieged, and so close girt about, that there can be no message sent in; . . . And so on the contrary, may the Besiegers make as good use of this Instrument to threaten and discourage the Besieged, . . . In case of great Fires, where usually all people are in a hurry, . . . In case a number of

Thieves and Robbers attaque a House that is lonely, and far from Neighbours, by such an Instrument as this, may all the Dwellers around about, within the compass of a Mile or more, be immediately informed, upon whose House such an attaque is made, the number of Thieves or Robbers, how armed and equiped, what manner of persons, with the colour and fashion of their Habits, and by what way they have made their escape, . . . or which way to pursue them."[91a]

The experiments Morland described were considerably more restrained than his accounts of the trumpet's "Manifold Uses." He worked first with a "*parabolical* Concave of fine Pewter, . . . and found that in the very same *Focus,* where the Rays of the Sun were so united, that in a minute of time they set on Fire a Deal Board, was the Voice of a man speaking near it, sensibly magnified." He appears also to have been the first to investigate experimentally the directivity of a sound source. The only measure of sound intensity available to him was his judgment of loudness, but after studying the sound distribution along the axis in front of and even into the mouth of his horn, and along a line at right angles to the axis, he concluded, quite correctly, "that the point where the Voice is most of all magnified or multiplied, is [at the center of the mouth]."[91b]

Morland transmitted his brochure to the Royal Society of London and to the *Académie des Sciences* in Paris, adding with becoming modesty that he "doubted not but this invention may be much improved." Each society duly reported his communication in its journal, and suggestions for improvement were indeed brought forth without delay. Several of these were concerned with a revival of the notion of using the horn as an aid to hearing and the analogy between spectacles and the ear trumpet. Like many good ideas, this one had to be rediscovered more than once before it could be effectively implemented.

The famous Ear of Dionysius, near Syracuse in Sicily, was probably the first and surely the largest "ear trumpet."[92] Leonardo da Vinci's simplified form of ear trumpet for listening to underwater sounds, his "long tube in the water," has already been mentioned,[93] and Giambattista della Porta (1538–1615) claimed that Adrianus (Publius Aelius Hadrianus, or Hadrian [76–138 A.D.]), consul of Rome, who had "this sense [of hearing] hurt, made hollow catches to hear better by; and these he fastened to his ears, looking forward." Porta went on to consider first those animals "that have the quickest hearing. . . . For this is confirmed in the Principles of Natural Philosophy, that when any new things are to be invented, Nature must be searched and followed." In this way, he soon led himself to the conclusion "that the Form of the Instrument for hearing, be large, hollow, and open, and with screws inwardly. For the first, if the sound should come in directly, it would hurt the sence; for the second, the voice coming in by windings, is beaten by the turnings in the ears, and is thereby multiplied, as we can see in an Eccho."[94] Perhaps the second part of this explanation was the basis for Kircher's conclusion that the cochleate form of ear trumpet is best.[89i, 90f]

Sir Francis Bacon restated in his posthumous *Sylva Sylvarum* (1627) the analogy Porta had drawn between spectacles and the ear trumpet. He did not acknowledge Porta as the source of this idea, although many of his other proposals were obviously drawn from Porta's *Magiae Naturalis*. He made his proposal more specific than Porta's, however, when he suggested that

> it be tried, for the help of the hearing, (and I conceive it likely to succeed) to make an instrument like a tunnel; the narrow part whereof may be of the bigness of the hole of the ear; and the broader end much larger, like a bell at the skirts; and the length half a foot more. And let the narrow end of it be set close to the ear: and mark whether any sound abroad in the open air, will not be heard distinctly from further distance than without that instrument; being (as it were) an ear-spectacle. And I have heard there is in Spain an instrument in use to be set to the ear, that helpeth somewhat those that are thick of hearing.[95]

Bacon did not explain why he thought this instrument was especially "likely to succeed," but it certainly would have done so, and it is regrettable that Bacon didn't go ahead and try to make one instead of just proposing that it be tried.

Presumably as unaware of the foregoing precedents as of the later suggestions of Kircher and Morland, Narcissus Marsh (1638–1713), then lord bishop of Ferns in Ireland, presented a "Doctrine of Sounds" to the Royal Dublin Society in which he too drew the analogy between seeing aids and hearing aids. He added, in language as quaint as Morland's, that such improvement of hearing acuity would be useful "in time of war, for discovering the enemy at a good distance, when he lies encamped behind a mountain or wood, or any such place of shelter which hinder the sight."[96]

Announcement of Morland's work at a meeting of the Paris Academy brought an immediate response from N. Cassegrain (late seventeenth century), variously referred to as a sculptor or as a professor of Chartres and otherwise better known through the association of his name with the reflecting telescope configuration now ascribed to the Scot James Gregory (1638–75). Cassegrain was a few years ahead of Marsh in reviving the eye-ear analogy, and he added a suggestion that the horn be given a hyperbolic profile.[97] Several of the subsequent meetings of the Paris Academy were devoted to lively discussion of the construction and use of such acoustical instruments and the relative merits of the proposals made by Cassegrain, Kircher, and Morland.

The first improvement of the Kircher-Morland horn that had survival value was suggested by (Sir [1693]) John Conyers (1644–1719), who demonstrated very impressively to the Royal Society in 1678 a modified form of horn that was much shorter and therefore easier to handle. Conyers did not use the terms *folded* or *reentrant*, but the drawings of his modification clearly disclose the modern form of reentrant folded horn with side feed.[98]

The modern exponential horn has its antecedent in the

demonstration by Richard Helsham (1680–1738) that "the best form for such tubes . . . is generated by the revolution of a logarithmic curve round its axis, as in a tube of this form the elastic bodies will increase in such a manner as most to increase the quantity of motion."[99] In analyzing this problem, Helsham suggested that the air in the horn be subdivided into thin laminae perpendicular to the axis (in much the same way that one now proceeds in setting up the differential equation of motion), and he then considered the transmission of motion from one lamina to the next. The conclusion he reached, that the area of any one lamina should be the geometric mean of the areas of its two neighbors, still constitutes a valid basis for arguing the superiority of the exponential form of horn profile.

The reasoning behind Helsham's proposal was soon discussed at some length by the Irish bishop of Clonfert, Mathew Young (1750–1800), with an engaging mixture of straight and circular logic. Young rejected Helsham's conclusions on the grounds that "though the motion of the whole mass may be increased, yet, the motion of that particular cylinder [of air] which produces the impulse on the ear, may be, and really is, diminished."[100] He concluded, nevertheless, that "These tubes should increase in diameter from the mouth-piece, because the parts, vibrating in directions perpendicular to the surface, will conspire in impelling forward the particles of air, and consequently, by increasing their velocity, will increase the intensity of the sound."

Young seems to have taken the idea that the wall of the horn itself vibrates from an argument he attributed to Kircher, to the effect that "the augmentation of sound depends on its reflection from the tremulous sides of the tube; which reflections, conspiring in propagating the pulses in the same direction, must increase its intensity." This may be what Young thought Kircher meant to say, but what he did say was even less cogent. "A sound in the narrow mouthpiece of the [conical] tube," Kircher wrote, "with the aid of the motion of the shut-in air, makes the whole tube tremble at once, and from the reciprocal reflexions which rebound against opposite points of the tube stimulated to reflecting by the said tremor, there arises that vigorous multiplication of the sounds."[89g]

Young might have come closer to the truth if he had concentrated on Kircher's earlier explanation of how sound is increased when propagated through a tube: "Not being able to escape it strives to recover by its longitudinal propagation what it would have lost if diffused by dispersal in a free medium. In these narrow confines being repeatedly reflected and multiplied it gains tremendous increase."[89j] And Young would have done still better to have devoted himself to the implications of the passage he quoted from Newton, according to which "it plainly appears how it comes to pass that sounds are so mightily increased in speaking-trumpets; for all reciprocal motion tends to be increased by the generating cause at each return. And in tubes hindering the dilation of the sounds, the motion decays more slowly, and recurs more forcibly; and therefore is the more increased by the new motion impressed at each return."[101a]

Each of these "explanations" came perilously close to claiming for the passive horn the nonpassive role of amplifying the total sound energy, but this can perhaps be excused since there did not yet exist any stated principle of energy conservation that would immediately brand such a notion as an overstatement. It is now known, of course, that the role of the horn is twofold: it *loads* the source more heavily, thus inducing the source actually to emit more sound energy than it could without the horn, and it *directs* more effectively the sound energy that is delivered by the source. Newton was already flirting with one of these ideas and Morland with the other, but real understanding did not come until the late nineteenth and early twentieth centuries, when Rayleigh dealt generally with the problem of source loading and Webster considered explicitly the theory of horns. In the meantime, the horn itself, as though with a fine disdain for whether it was understood or not, continued to be what Morland had claimed, an acoustical "Instrument of Excellent Use."

Refraction, Diffraction, and Interference

The optical phenomena of refraction, diffraction, and interference were first elucidated during the seventeenth century and each was either at once or eventually recognized to be as important for sound as for light. Observers since antiquity had concerned themselves with the deviation, or refraction, of light rays as they enter or emerge from a transparent medium such as water or glass. Willebrord Snel (or Snellius; often, but erroneously, Snell) (1591–1626) prepared an essay in 1620 that finally unraveled this ancient puzzle, although he seems to have been unaccountably negligent about publishing it. As a consequence, Descartes's followers questioned its existence and challenged Snel's claim to the discovery. Huygens saw it in manuscript, however, and made use of it later in his own writings; and it is largely on this evidence that Snel's claim is secured.

Snel's law of refraction, although not yet expressed in its modern form as a ratio of sines, was first published[102] in 1637 by René Descartes (1596–1650). Whereas Snel had established his form of the law of refraction on the basis of experiment, Descartes deduced it independently as a theoretical consequence of three assumptions, two of which turned out to be wrong. Pierre de Fermat (1601–65) voiced proper objections to Descartes's assumptions and succeeded in deriving the same law by making the quite different assumption that light always travels from the source point in one medium to a receiving point in the second medium by whatever path will take the least time.[103] This assumption is now referred to as "Fermat's principle," but it can easily be identified as a minor modification of Aristotle's *lex parsimoniae*, which had been widely invoked by many others before Fermat, for example Heron, Grosseteste, Copernicus, Kepler, et al.[104]

These three independent solutions of the problem of refraction afford, incidentally, an illustration of the potency of adroit formulation of a physical problem. Everyone who had considered the phenomenon of refraction before Snel took it up had concentrated his attention on the angle through which the ray was bent, a small angle that is still difficult to express simply in terms of the propagation velocities in the two media. Snel's greatest contribution to the solution of this problem was to shift the point of view so that attention was concentrated on the relation between the angles which the incident and refracted rays make with the normal at the refracting surface.

The optical phenomenon called diffraction was observed for the first time in the careful experiments of Francesco Maria Grimaldi (1618-63), a professor of mathematics in the Jesuit College at Bologna. These were described in a posthumous publication of 1665 which showed that light does not always travel in straight lines but can bend slightly around corners.[105] Newton, Hooke, and Christiaan Huygens (or Huyghens; 1629-95) all repeated these experiments, and before long the new phenomenon became one of the two or three central issues in the lively contest between the wave and corpuscular theories of light. The modern resolution of this dilemma, of course, is to embrace both its horns and to invoke the machinery of wave interference and diffraction to predict the paths or positions of either corpuscular photons or material particles; but it took all of the two centuries following Newton's time to bring into being the elements of this compromise and a philosophy that could accept it.

Newton's stout adherence to the corpuscular viewpoint led him to advance a strained doctrine based on "Fits of easy Reflexion" and "Fits of easy Transmission" in order to explain the colors seen in thin films,[106a] and it also led him into an equally untenable position with regard to diffraction. He argued against the possibility that light could "consist in Pression or Motion, propagated through a fluid Medium," because, he said, in that case, "it would bend into the Shadow." This is just what does occur in diffraction, of course, but Newton was led astray by the quantitative aspects of the phenomenon that stem from the disparity in the wavelengths of sound and light. He conceded that "Rays which pass very near to the edges of any Body, are bent a little by the action of the Body,"[106b] but "So soon as the Ray is past the Body, it goes right on." He had observed this "going right on" in his diffraction experiments and this led him to reject the wave hypothesis:

> For Pression or Motion can-not be propagated in a Fluid in right lines, beyond an Obstacle which stops part of the Motion, but will bend and spread every way into the quiescent Medium which lies beyond the Obstacle. . . . The Waves on the Surface of stagnating Water, passing by the sides of a broad Obstacle which stops part of them, bend afterwards and dilate themselves gradually into the quiet Water behind the Obstacle. The Waves, Pulses or Vibrations of the Air, wherein Sounds consist, bend manifestly, though not so much as the Waves of Water. For a Bell or a Cannon may be heard beyond a Hill which intercepts the sight of the sounding Body, and Sounds are propagated as readily through crooked Pipes as through streight ones.[106c]

These remarks, which Newton cited in the negative, were influential in suppressing the wave theory during most of the eighteenth century, but they came to its support after Thomas Young (1773-1829) had revived and well-nigh clinched the wave theory with his announcement of the principle of interference in 1802.[107] Young drew on acoustical analogy to explain the black spot in the center of Newton's rings, and it can be inferred that he was led to consider these optical problems by the close analogy he had previously drawn between sound and light and by the similarity between optical interference and the phenomenon of beats between two sounds differing only slightly in frequency.[108] Young was castigated for "attacking" Newton's theories in spite of his generous acknowledgment of debt to the wave concepts in Newton's *Opticks.*

Reinforcements for the wave theory were soon brought up by Augustin Jean Fresnel (1788-1827), a young Royalist military engineer who turned his hand to experimental science after Napoleon's second exile. Without being aware of Young's earlier disclosure, Fresnel announced to the French Academy in 1818 his own independent discovery of the principle of interference.[109] Fresnel's analysis of interference and diffraction was mathematical and not only gave strong support to the wave theory itself but also explained the differences between the diffraction of sound and of light in terms of the differences in wavelength. This was, of course, just what was needed to salvage Newton's citation of acoustical diffraction, and it was soon recognized that diffraction is of even greater importance in acoustics than it is in optics, owing to the similarity in size between common diffracting objects and the wavelengths of the sounds of ordinary experience.

In his analysis of diffraction, Fresnel had made extensive use of "Huygens' principle," whereby successive positions of a wavefront are determined by the envelope of secondary wavelets. Huygens had set forth this scheme in his classic *Traité de la lumière,*[110] where he described it as follows: "each particle of the matter in which a wave proceeds not only communicates its motion to the next particle to it . . . but . . . also gives a motion to all the others which touch it. . . . The result is that around each particle there arises a wave of which this particle is the center. . . . the particular waves produced by the particle . . . never conspire at the same instant to make up a wave . . . except precisely in the circumference . . . which is their common tangent."[111]

This elegant conception has survived with less change than almost any other analytical scheme proposed during the seventeenth century and it is still widely used in the study of all types of wave propagation. Huygens was well aware of its specific applicability to sound waves, since he held that light itself was a similar kind of longitudinal wave motion. But Huygens's claim to remembrance does not rest alone on this justly celebrated "principle." He, like Leonardo and Galileo, had been blessed with the combination of mathematical skill and practical ingenuity that enabled him to play on the continent a role closely paralleling that of Newton in England.

The blind but still alert Galileo, shortly before he died, had dictated to his pupil Viviani the description of a scheme for making a pendulum-controlled clock. Unfortunately neither Viviani nor Galileo's son Vincenzio carried out this plan (the claim advanced in the *Saggi* of the Accademia del Cimento notwithstanding). It remained for Huygens, working independently, to reduce this important invention to practice in 1567, as was duly attested by a royal patent issued to him that year.

Huygens's continuing studies of pendular and rotary motion matured in the definitive publication *Horologium Oscillatorium* [1673], in which he anticipated much of the content of Newton's laws of motion. For example, Huygens discussed correctly the law of centrifugal force, and he was the first to introduce and define the moment of inertia and the concept of conjugate centers of oscillation for a solid body. He seems also to have appreciated the distinction between mass and weight; but although he suggested that a pendulum could be used to establish the ratio of mass to weight, he never quite brought himself to accept the full implications of Newton's law of universal gravitation.[112a] Finally, Huygens gave us the first statement of the important principle of energy conservation in mechanical systems[112b] in his report of an earlier study of collision undertaken jointly with John Wallis and Sir Christopher Wren (1632–1723) at the request of the Royal Society.

The history of science is spotted, unfortunately, with examples of the neglect of a scientist's duty to make available by publication a permanent record of his work. Leonardo,

Snel, and Fermat can be cited for antivirtuosity in this respect; but they were not the only ones, and surely not the last, to be guilty of such sins of omission. The preface to Huygens's *Treatise on Light*, in Thompson's graceful English translation, is well calculated to enlist the sympathy of such dilatory scientists addicted to delay, and the concluding judgment of the following passage can still be held up both as shining example and worthy precept. Huygens began,

> One may ask why I have so long delayed to bring this work to light. The reason is that I wrote it rather carelessly . . . with the intention of translating it [into Latin, of course] . . . in order to obtain greater attention to the thing. . . . But the pleasure of novelty being passed, I have put off from time to time the execution of this design, and I know not when I shall ever come to the end of it, being often turned aside either by business or by some new study. Considering what I have finally judged that it was better worth while to publish this writing, such as it is, than to let it run the risk, by waiting longer, of remaining lost. [N. 110]

Precursors of Instrumentation: Stethoscope and Siren

Two acoustical instruments that were to be exploited in the nineteenth century, the stethoscope and the rotating toothed wheel for tone production, had their origins in the seventeenth-century work of Robert Hooke. Boyle recognized Hooke's flair for experimentation during their joint labors with Guericke's vacuum pump and later secured for Hooke the post of Curator of Experiments for the Royal Society. Hooke was frequently embroiled in conflicts over the assignment of credit for inventions and discoveries, but he seems to have thrived on it. Newton, on the other hand, always avoided controversy when he could, and it was said that Hooke's contentiousness, plus his advocacy of the wave theory of light, influenced Newton to delay the publication of his *Opticks* until a year after Hooke's death in 1703.

Hooke is best remembered for his discovery of the linear relation between the applied force and the extension of a spring ("Hooke's law"),[113] which he claimed to have discovered in 1660. He delayed its publication for eighteen years, however, "designing to apply it to some particular use." The "use" to which he referred was his invention of the hair-spring-controlled balance wheel for watch or clock escapements. He took much pride in this invention, which competed on nearly even terms with Huygens's pendulum as a replacement for the unreliable and uncontrolled oscillating *foliot* of earlier timepieces.

Hooke never published an adequate description of his experiments on the production of sound by a rotating toothed wheel, but his approach to the problem is indicated by the entry in his diary for 15 January 1675/6, where he gave his "notion of sound, that it was nothing but strokes within a Determinate degree of velocity," and said that he "would make all tunes by strokes of a hammer"; and that "scraping the teeth of a comb, the turning of a watch wheel, &c., make sound."[114a] Some two months later, Hooke "Directed Thompion [a famous London clock maker] about sound wheels. The number of teeth."[114b] He seems thus to have worked toward the traditional toothed-wheel experiment about which Waller, his biographer, testified as follows: "In July the same year [1681] he shew'd a way of making *Musical and other Sounds*, by the striking of the Teeth of several Brass wheels, proportionally cut as to their numbers, and turned very fast round, in which it was observable, that the equal or proportional stroaks of the Teeth, that is, 2 to 1, 4 to 3, & c. made the Musical Notes, but the unequal stroaks of the Teeth more answer'd the sound of the Voice in speaking."[115] Felix Savart (1791–1841) revived this scheme in 1830 and used it to measure the exact frequency corresponding to particular musical notes. Savart used first a brass wheel 24 cm. in diameter, bearing 360 teeth on its periphery, and then progressed to a king-size 82 cm. wheel with 720 teeth, which must have been very impressive in operation. One of Savart's experiments with this large wheel furnished, incidentally, the first direct evidence bearing on the upper frequency limit of audibility, "the sound being still perceptible even when the number of tooth impacts was 24,000 per second."[116]

Hooke's second anticipation of nineteenth-century acoustical instrumentation dealt with the stethoscope, which was definitively established in medical usage in 1819 by René Théophile Hyacinthe Laennec (1781–1826), a French physician and professor of the Collège de France. Hooke observed that:

> The Sense of Hearing does not altogether so much instruct as to the Nature of things as the Eye, though there are many Helps that this Sense would afford by a greater Improvement, there may be a Possibility that by Otocousticons many Sounds . . . may be made sensible. . . . There may be also a Possibility of discovering . . . the Motions of the Internal Parts of Bodies, whether Animal, Vegetable, or Mineral, by the sound they make, that one may discover the Works perform'd in the several Offices and Shops of a Man's Body, and thereby discover what Instrument or Engine is out of order, what Works are going on at several Times, and lies still at others, and the like; . . . I have this Incouragement, not to think all these things utterly impossible . . . so the believing them possible may perhaps be an occasion of taking notice of such things as another would pass by without regard as useless. And somewhat more of Incouragement I have also from Experience, that I have been able to hear very plainly the beating of a Man's heart, and 'tis common to hear the Motion of Wind to and fro in the Guts, . . . so as to their [sounds] becoming sensible they require either that their Motions be increased or that the Organ be made more nice and powerful to sensate and distinguish them (to try the Contrivance about an Artificial Timpanum) as they are, for the doing of both of which I think is not impossible but that in many cases there may be Helps found, some of which I may as Opportunity is offer'd make Tryal of, which if successful and useful, I shall not conceal.[117]

Just what Hooke meant by an "Otocousticon" or by a "Contrivance about an Artificial Timpanum" is *not* explained here or elsewhere; and although Hooke appears to have been the first to discuss "Helps" for the audition of heartbeats, he was almost certainly not the first to have heard them. William Harvey (1578–1657), in his classic treatise of 1628 on the

circulation of blood, had drawn a suggestive analogy in referring to these sounds: "It is easy to see when a horse drinks
that water is drawn in and passed to the stomach with each
gulp, the movement making a sound, and the pulsation may
be heard and felt. So it is with each movement of the heart
when a portion of the blood is transferred from the veins to
the arteries, that a pulse is made which may be heard in the
chest."[118] One medical historian has suggested that Hippoc-
rates, Erasistratus, and Galen may also have been aware of
these heart sounds, but the evidence for this is not strong,
and it can only be said that these were three physicians who
should have heard them. Even Harvey's strong testimony was
challenged in some quarters, presumably by those who pre-
ferred disputation to direct trial, and one Venetian doctor
suggested snidely that perhaps such heart sounds were "only
to be heard in London."[119]

In spite of Hooke's prophetic suggestion that one might
discover what engine was out of order in a man's body by the
sound it makes, the diagnostic value of heart sounds was almost entirely ignored by the physicians who studied the heart
during the century and a half following Harvey's great work.
All this changed abruptly when it became apparent that
Laennec's stethoscope was the wanted "Help" to which
Hooke had referred so long before. Laennec described the
circumstances which led him to this important invention
with touching directness. "In 1816," he wrote

> I was consulted by a young woman labouring under general symp-
> toms of diseased heart, and in whose case percussion and the applica-
> tion of the hand were of little avail on account of the great degree of
> fatness. The other method just mentioned [application of the ear to
> the precordial region] being rendered inadmissible by the age and sex
> of the patient, I happened to recollect a simple and well-known fact
> in acoustics, and fancied, at the same time, that it might be turned
> to some use on the present occasion. The fact I allude to is the aug-
> mented impression of sound when conveyed through certain solid
> bodies,—as when we hear the scratch of a pin at one end of a piece of
> wood, on applying our ear to the other. Immediately, on this sugges-

tion, I rolled a quire of paper into a sort of cylinder and applied one end of it to the region of the heart and the other end to my ear, and was not a little surprised and pleased, to find that I could thereby perceive the action of the heart in a manner much more clear and distinct than I had ever been able to do by the immediate application of the ear. From this moment I imagined that the circumstances might furnish means for enabling us to ascertain the character, not only of the action of the heart, but of every species of sound produced by the motion of all the thoracic viscera.[120]

Laennec's cautious, yet triumphant, confidence in the diagnostic merit of auscultation, or what Hooke called listening to "the Works perform'd in the several Offices and Shops of a Man's Body," has turned out to be thoroughly justified, and few contributions to medical instrumentation have survived as long and with so little change as the stethoscope.

Boundaries Away

The chronological boundary afforded by Newton's work has been overstepped at several places in the foregoing in order to preserve the thread of continuity running through some of the fields of inquiry. On the other hand, some experiments performed before or during Newton's lifetime have been reserved for [the Appendix],* where they can be juxtaposed with the theoretical developments they supported or stimulated. It is noteworthy, however, that almost all the acoustical achievements in the sixteenth and seventeenth centuries so far remarked were made without recourse to, and without substantial assistance from, parallel advances in the science of mathematics. From this point of view, Newton's time was indeed the culmination of an age of experiment—an age that explored, confirmed, and established the validity of orderly procedure from hypothesis to experiment and theory.

It is hardly necessary to add that the spirit of free inquiry

Editor's note: In his manuscript, Frederick Hunt referred here to chapter 3. The finished portion of this chapter is now published as an appendix.

and experiment was not wholly confined to science during
the sixteenth and seventeenth centuries. The contagious
spirit of the Renaissance provided a stimulus that dared men
everywhere to question, and to seek and even to fight for,
freedoms. Among the by-products of this ferment were the
great voyages of discovery: Columbus to the New World
(1492), Vasco da Gama around the Cape to India (1498),
and Magellan around the globe itself (1519-22). Out of the
ferment also came to Europe the travail of the Reformation
(1517-?) and continuing conflicts on every hand, until the
Treaty of Westphalia (1648) finally established a balance of
power and a temporary period of peace.

HARMONIE
VNIVERSELLE

Nam & ego confitebor tibi in vafis pfalmi veritatē tuam:
Deus pfallam tibi in Cithara, fanctus Ifrael. *Pfalme 70.*

Title page from Marin Mersenne, *Harmonie Universelle*, 1636. Courtesy of
the John Herrick Jackson Music Library, Yale University.

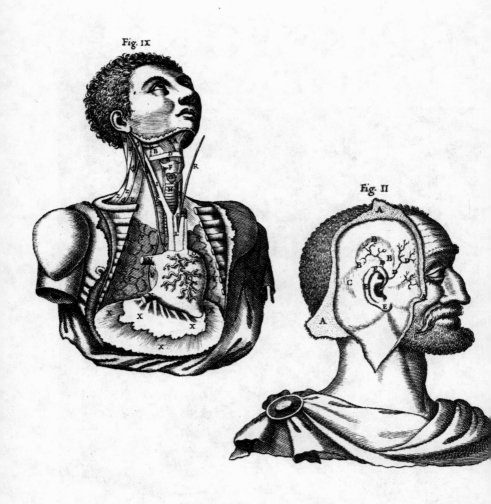

The production (*above*) and reception (*below*) of sound. From Athanasius Kircher, *Musurgia Universalis*, volume 1, 1650. Courtesy of the Beinecke Rare Book and Manuscript Library, Yale University.

The propagation of sound: reflection and echoes. From Athanasius Kircher, *Musurgia Universalis*, volume 2, 1650. Courtesy of the Beinecke Rare Book and Manuscript Library, Yale University.

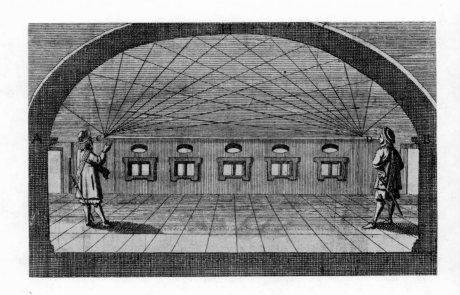

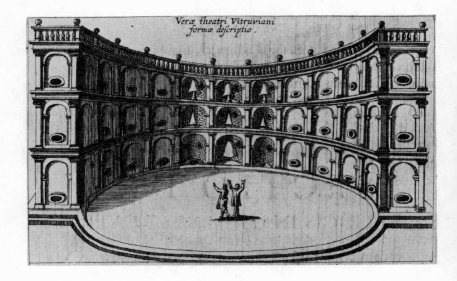

Veræ theatri Vitruviani formæ descriptio.

Architectural acoustics, indoors and outdoors: (*above*) the focusing of sound by curved surfaces; (*below*) the theater of Vitruvius. From Athanasius Kircher, *Phonurgia Nova*, 1673. Courtesy of the John Herrick Jackson Music Library, Yale University.

Horns for use by the military to summon troops. From Athanasius Kircher, *Phonurgia Nova*, 1673. Courtesy of the John Herrick Jackson Music Library, Yale University.

Horns to be used in residences, for paging, music, or eavesdropping. From Athanasius Kircher, *Phonurgia Nova*, 1673. Courtesy of the John Herrick Jackson Music Library, Yale University. It is not clear that such horns were ever built, although large horns have been built in the twentieth century for sound reproduction.

APPENDIX

The two centuries between Newton's time and that of Rayleigh embrace a remarkable period of development and growth in the parts of physical science now designated as "classical." The calculus was invented and developed as a powerful tool of analysis, and was used to deal exhaustively with the problems of analytical mechanics; heat became an understandable mode of motion and the laws of thermodynamics a comprehensive account of all energy-exchange phenomena; and electricity evolved from pith balls and amber to a theory and practice of electromagnetism capable of sustaining a major industry. Science historians have dealt fondly and extensively with these major triumphs of human understanding. With depressing uniformity, however, these same historians found remarkably little to say about sound during the same period, their silence tacitly endorsing the conclusion Chladni drew in 1802, "that the science of acoustics has been more neglected than most other portions of physics."[1] But with the advantage of hindsight, this very silence, like "the curious incident of the dog in the night-time"[2] remarked by Sherlock Holmes, is enough to arouse rather than to suppress a lively interest in accounting for the origins of the modern science of physical acoustics in the eighteenth and nineteenth centuries.

In the seventeenth century the study of all branches of physical science was predominantly phenomenological. Falling bodies, swinging pendulums, vibrating strings, and the motion of the planets were for Galileo problems of comparable stature to be attacked by similar methods of logical inference based on observation. This proved to be a transient state of parity, however, as one after another of the scientific disciplines moved into the spotlight of attention during the next two centuries. It was inevitable that the science of mathematics should have led this parade since, as Bacon had already explained, "many parts of nature can neither be invented with sufficient subtlety, nor demonstrated with sufficient perspicuity, nor accommodated to use with sufficient dexterity, without the aid and intervention of Mathematic."[3] It was no less inevitable, although less obvious, that mechanics, heat, and electricity had to come of age first before any comparable progress in understanding could be made in the field of sound.

Such a time lag in the development of the science of sound has its origin in the nature of sound itself. In particular, two definitive characteristics of the modern science of physical acoustics have special relevance to its pace of evolution since Newton's time. In the first place, the formulation of an analytical theory of sound demanded (and still demands) wholesale intellectual borrowing across the mounting barriers between such diverse disciplines as fluid mechanics, thermodynamics, elasticity, and electromagnetics. And since one cannot borrow until the lender has something to lend, the growth of physical acoustics had to wait until each of these specialties had matured in its own right.

The second controlling factor in the growth of the science of acoustics was the almost total inaccessibility of quantitative information about the intensive aspects of sound phenomena. The sound waves of ordinary conversational experience represent rapid alternating variations of pressure usually amounting to less than a millionth of the equilibrium atmospheric pressure. Viviani in Italy, Mariotte in France, and Boyle in England could and did perform experiments "touching the spring of the air" that were meaningful with regard to the equilibrium atmospheric pressure and its gross variations; but the possibility of bringing the minute variational pressures or variational velocities involved in sound waves under accurate observation was so remote that it was hardly attempted until the mid-nineteenth century and not satisfactorily achieved until the twentieth. As a result, what Bacon had said of an earlier period remained stubbornly true, and only the qualitative aspects of sound that could be forwarded by subjective observation could be "inquired" in any way except superficially. The retarding effect on acoustics of this lack of a hard core of experimental fact can hardly be over-emphasized, for as Lord Kelvin was to say, "when you can measure what you are speaking about and express it in numbers, you know something about it; but when you cannot measure it . . . your knowledge is of a meagre and unsatisfactory kind."[4]

The Birth of the Calculus

The theoretical development of acoustics, no less than the experimental, was influenced by the small-signal feature of sound waves that leads to their characterization as a variational aspect of fluid mechanics. Present-day students armed with the calculus are taught that it is of little consequence whether one starts with a curve and differentiates to find the slope or starts with the slope and integrates to find the curve. No such choice of concept was available, however, to Boyle or Hooke or Pascal, as perforce they addressed themselves first to the challenges in the static equilibrium of linear stress and strain, and in the problems of hydrostatic pressure. In a voice scarcely heard in the seventeenth century, the variational aspects of mechanics in the small, no less than the planets in their courses, fairly clamored for the infinitesimal calculus. Fortunately, Newton at least was listening to the planets.

It may well be, as one theory of historical evolution would have it, that the discoveries of the calculus and of universal gravitation were nascent in the seventeenth century, and that these concepts would surely have asserted themselves even if they had not been formulated by Sir Isaac Newton (1642–1727). There are many antecedents of the infinitesimal calculus in the writings of Stevin, Kepler, Cavalieri, Fermat, and Wallis, to mention only a few, which might constitute circumstantial support for such a hypothesis.[5] There were similar precursors for Newton's laws of motion and the law of universal gravitation. Leonardo and Galileo had each disclosed one or more of the laws of motion and other writers had conceived of the heavenly bodies moving under the influence of a force proportional to the inverse square of the distance. For example, Mersenne, Huygens, Hooke, Halley, and Sir Christopher Wren (whose later fame as an architect overshadowed his considerable distinction as a mathematician), each independently had suggested, with varying degrees of conviction, that if Kepler's harmonic law was correct the force law would be an inverse square.

By the time he was twenty-six years old, Newton also had deduced the appropriateness of an inverse square law for gravitating point masses, but he deduced as well the important additional conclusions that the orbits were ellipses and that the sun and earth exerted their attractions from approximately the location of their centers. These results were obtained during the two-year period Newton spent at his home in Grantham, Lincolnshire, while Cambridge University was closed and London was besieged by the Great Plague of 1665. This was an immensely fruitful period for science, inasmuch as within these same two years Newton also rounded out his conception of a calculus of "fluxions" and performed his famous prism experiments on the decomposition of white light into its color constituents. The latter optical experiments formed the subject matter of Newton's first published paper,[6] but he did not bother to make his other advances in mathematics and mechanics known except through his correspondence with friends and his poorly attended lectures at

Cambridge, to which he returned as Lucasian Professor in 1669.

It can be added that the rousing contention set off by the publication of Newton's optical experiments regrettably fortified his natural disinclination toward publication. Thus, it was only at the earnest entreaty of Sir Edmund Halley (1656–1742) that Newton was persuaded to take up again in 1685 the task of summarizing his work on mechanics and gravitation. The one lacuna in his earlier work on gravitation had been a proof of the theorem that spheres interact *exactly* as if their masses were concentrated in their centers; but Newton soon filled in this gap and within seven months he had completed the first book, and within eighteen months the entire manuscript, of his monumental *Philosophiae Naturalis Principia Mathematica* (see pt. 2, n. 101). This was quickly printed at Halley's expense and was published in 1687 under the imprimatur of Samuel Pepys, then president of the Royal Society.

Most of the demonstrations in the *Principia* were couched in terms of classical geometry, in the use of which Newton displayed a virtuosity unmatched either before or since. These geometrical expositions placed a heavy burden on most readers, then no less than now; but the *Principia* was a new landfall in man's quest for an understanding of nature, and almost no scientist who came after Newton could claim to have escaped its direct or indirect influence. Some of his contemporaries ignored and some sought to controvert his "system of the world," but later generations universally acclaimed it. Laplace conceded that "it was reserved for Newton to make known to us the general principle of celestial movements," to which he added, "Nature while endowing him with a profound genius, took care also to place him in the most favorable circumstances."[7] Lagrange, "who often referred to him [Newton] as the greatest genius who had ever lived," took note also of his timeliness by adding, "and the most fortunate; one finds only once a system of the world to establish."[8] Newton himself had remarked, in a letter to Hooke, that "if I have seen farther, it is by standing on the shoulders of giants."[9] This quotation is often cited as proof of Newton's humility, but he was well aware of his own stature and it is more likely that he was taking this opportunity to remind Hooke that he *had* seen farther and that he *did* tower above the contemporary giants.

When discussing the genesis of the calculus, the name of Gottfried Wilhelm Leibniz (1646–1716) must always be mentioned in the same breath with that of Newton. A bitter and acrimonious dispute over priority distracted both Newton and Leibniz, as well as followers in the camp of each, for many years. The balance of evidence indicates now that Newton certainly "had it" first but did not publish it until much later; whereas Leibniz "got it" later but published it first. What may be more important, as far as later generations are concerned, is that Leibniz expressed his ideas in a form and notation so much superior to Newton's that it is almost entirely Leibniz's *language* of the calculus which survives in modern usage.

There was credit enough for both, however, since there can be little doubt that the invention of the calculus was one of the three crowning scientific achievements of the seventeenth century. The other two, of course, were the laws of motion and of universal gravitation—and Newton's name is attached to all three! The influence of the several mathematical inventions commonly lumped together as the "calculus" was so great, in fact, that most physical scientists as well as mathematicians were chiefly preoccupied throughout the first half of the eighteenth century with the development and refinement of these new mathematical tools. Only toward the middle and during the second half of the eighteenth century did the focus of attention of the best scientific minds of the period shift to applications of the new mathematics to physical problems.

The course of development of these mathematical tools

was not uneventful. The followers of Newton and of Leibniz made up two dissident groups, one in England and one on the Continent, and the argument about who invented the calculus divided them so sharply that each was prevented from reaping the compounded benefits that accrue from free intercommunication. Among the members of the English school were several men whose names are still attached to theorems or methods that become routinely familiar to present-day students of intermediate calculus: Brook Taylor (1685–1731), Abraham de Moivre (1667–1754), Thomas Simpson (1710–61), and Colin MacLaurin (1698–1746). MacLaurin was probably the ablest of these, but he is less enviably marked as having retarded the development of mathematics in England for the next two generations through his stubborn cleavage to Newton's methods and his refusal to adopt the more powerful analytical formulation of the calculus that was by then gathering momentum on the Continent. As a result, the center of gravity of research in the area of natural philosophy gradually shifted to the Continent during the eighteenth century.

Two of the most influential followers of Leibniz were the elder Bernoulli brothers, James (1654–1705) and Johann (1667–1748). The latter appears to have been less than upright and generous in some of his personal dealings, even with his own son, but he seems also to have had the precious faculty of inspiring zeal in his students. These two Bernoullis, together with Leibniz and a few of their students, such as G. F. A. l'Hôpital (Marquis de St. Mesme) (1661–1704), Jacopo Francesco (Count) Riccati (1676–1754), and Gabriel Cramer (1704–52), succeeded in firmly establishing in Europe by the middle of the eighteenth century the basic methods and the modern language of analytical geometry and of the differential and integral calculus.

As these new conceptions matured, three great men appeared who gave definitive form to the mathematical tools and began to apply them to physical problems. They were

Leonhard Euler (1707–83), whose stature continues to grow as the accomplishments of the eighteenth century become more widely appreciated, Jean-le-Rond d'Alembert (1717–83), and Daniel Bernoulli (1700–83), the latter being the most distinguished of the three sons (all mathematicians!) of the elder Johann Bernoulli. And scarcely a generation or two later, five other giants came forth who further embellished these mathematical tools and put them to such splendid use as to make of the eighteenth century a golden age of mechanics: they were Joseph Louis Lagrange (1736–1813), Pierre Simon Laplace (1749–1827), Adrian Marie Legendre (1752–1833), Jean Baptiste Joseph Fourier (1760–1830), and Siméon Denis Poisson (1781–1840).

The Newtonian Acoustical Dilemma:
Theme and Resolution

The second Book of Newton's *Principia* dealt with the rational mechanics of fluids and marked the first serious attack on the frontiers of this field since the work of Stevin. The text was extensively revised as the *Principia* went through its second and third editions in 1713 and 1726, but even in its final form it lacks the definitive crispness of Books 1 and 3, in which "Newton's Laws" of mechanics and of gravitation are boldly set forth and for the most part correctly elaborated. It is not, therefore, the new and often shaky results he adduces that earn Newton credit in the field of fluid mechanics, but rather the originality of his approach and the fact that he tackled the problems at all.

In spite of his declared aversion to "framing hypotheses," Newton introduced several of them in Book 2, sometimes supposing the fluid to be a "rare" medium composed of small quiescent particles freely disposed at equal distances from one another, sometimes supposing these particles to flee one another under forces inversely as their distances, and at other times supposing the fluid to be a continuous "compressed" medium. The latter viewpoint, incidentally, provided

Newton with the vehicle for a further hypothesis concerning the linear dependence of the forces of viscosity "arising from the want of lubricity" on the velocity gradient; but the only use he made of this correct and original suggestion was to take it as the basis for demolishing the Cartesian concept of a heavenly firmament filled with planetary vortices.

Newton drew on first one and then another of these fluid models in reckoning the resistance to the motion of bodies, but most of the results were spurious and served chiefly to challenge following generations to find and correct his errors. For example, in his deductions based on a frictionless continuum model there appears no trace of the correct inference that the resistance should vanish for streamline or irrotational flow [later identified as "d'Alembert's Paradox"]. On the other hand, Newton's "corpuscular" fluid models did not include any accounting for interparticle collisions and hence could only be held valid for gases at least as highly rarefied as the atmosphere at very high altitudes. The mixed blessings of ballistic missiles and artificial satellites have kindled a modern interest in the resistance of such rare mediums, but as a constitutive hypothesis Newton's was much too restricted to serve adequately as a basis for explaining even the primitive behavior of liquids or gases at ordinary pressures. Newton did deduce for his corpuscular model with repellent particles a linear relation between density and pressure that is correct for rarefied or perfect gases, but his associated inquiries about the resistance of such a "rare" medium were aptly characterized by d'Alembert as "a research of pure curiosity, . . . not applicable to nature."[10]

Newton devoted section 8 of Book 2 of the *Principia* to "the motion propagated through fluids," and he gave there for the first time, in Propositions 47–50, an analytical formulation of the same mechanism of sound-wave propagation that had been described qualitatively with such prescience nineteen centuries earlier in Straton's *De Audibilibus* (see pt. 1, n. 38). Here, as elsewhere in the *Principia*, the geometrical mode of exposition intruded to cost Newton many readers and even more understanders. What he proved in these four propositions was that the forces acting on the "particles of the fluid" displaced from their positions of equilibrium behave like the restoring forces acting on a displaced pendulum, and that the motion of the particles followed the laws of the oscillating pendulum; that the speed at which sound "pulses" travel is independent of their "intensity" if the "contractions and dilatations" are not "exceedingly intense"; that the "breadth of one pulse" [i.e. the wavelength] is the ratio of its rate of advance to the frequency; and that the speed of propagation of sound pulses is given by the square root of the ratio of the "elastic force" to the density of the medium.

Newton specified the condensation almost adequately, but many of the terms he used, such as the sound "pulse," which is central to the argument and the "tremulous body," which generated it, remained undefined and too vague to serve either as solutions or boundary conditions for a proper wave equation. Nevertheless, he was on firm if somewhat restricted ground in his conclusions, up to the point at which he identified the "elastic force" of the medium with the mean atmospheric pressure, which he chose to specify in terms of "a height of uniform air, whose weight would be sufficient to compress our air to the density we find."

It is commonly said that Newton made the "mistake" of supposing that the sound-bearing medium would be compressed and expanded isothermally, but this is not correct and puts into Newton's mouth a concept that was still unknown in Newton's time. To be sure, what Newton did assume, namely, that the "elastic force" of the fluid is proportional to its "condensation," is *now* known to be equivalent to an isothermal hypothesis, but knowledge of this equivalence was a dearly won triumph of the nineteenth century. As a consequence of this "mistake," however, Newton's first calculation of the speed of sound gave the value 968 feet per second, which is too low by about 16 percent. He went on, with massive complacency, to say that the speed "is found by experiment" to lie between 866 and 1,272 feet per second,[11] and he concluded, therefore, that his theory was in satisfactory agreement with the phenomena!

Newton's complacency was soon dissipated when the sound speed experiments reported by Walker (see pt. 2, n. 61) in 1698 and by Derham (see pt. 2, n. 79) in 1708 firmly established the dilemma by showing that there was indeed a real discrepancy between theory and experiment. Thereafter, throughout most of the following century, the resolution of the sound-speed dilemma was a prime target for theorists in fluid mechanics, for experimentalists, and for speculators who sought to explain away the difficulty by importing additional and sometimes fanciful hypotheses about the medium. As an example of the latter, Derham, in his otherwise cautious report, raised the question "whether it was the air itself or certain ethereal, subtle, gaseous, or thick parts of the same which were the real vehicles of sound."

Newton himself succumbed to the temptation of tampering with the medium when he revised his calculations of sound speed in the second (1713) edition of the *Principia*. For the air itself he found this time a speed of 979 feet per second. Then, in a monstrous exhibition of teleological data manipulation, Newton proceeded to add to this speed 109 feet per second as an "allowance for the crassitude of the solid particles of the air, by which the sound is propagated instantaneously." After that he increased the sum of these in the ratio of 21 to 20 to compensate for "the vapors floating in the air being of another spring," and thus by such high-handed maneuvering he reached the conclusion that "sound will pass through 1,142 feet in one second of time," a value for the speed that just happened to coincide exactly with the one reported to the Royal Society five years earlier by Derham!

A few years later Jean-Jacques d'Ortous de Mairan (1678–1771) proposed to the Paris Academy (1719), and later published (1737), another ad hoc assumption about the acoustic properties of the medium. This took the form of a suggestion that the propagation of many different kinds of tones in the air at the same time would require that the air be composed of different parts of differing elasticities.[12] Mairan took this as a point of departure from which to elaborate extensively on a casual reference Newton had made to the similarity between the space distribution of the colors of the spectrum produced by his prism and the lengths of the segments of a string that would sound the notes of a musical scale.[13]

Newton's analogy between sound and color was soon picked up by the music historian Sir John Hawkins (1719–89), who found in it another demonstration of "what has often been asserted . . . that the principles of harmony are discoverable in so great a variety of instances, that they seem to pervade the universe."[14] The analogy itself was hardly less pervasive and it has been revived from time to time ever since by various people, among them more recently the physicist Albert Abraham Michelson (1852–1931).[15] With each renewed elaboration of the concept of "color music," Newton came more and more to be charged with a lapse into mysticism, and Sir J. J. Thompson went so far as to suggest that it was "the siren's song of these harmonics" that lured Newton into false conclusions regarding the continuity of the spectrum and the uniform dispersion of all substances.[16] The actual text of the *Opticks*, however, defends Newton quite adequately against the charge of mysticism, in this connection at least, and tends rather to establish that he was unusually perceptive to have recognized and identified a not very obvious numerical coincidence.

Euler's first publication, which appeared while he was still a young man barely twenty years old, was a brief nonmathematical dissertation on the physics of sound. In this tract, and in a companion paper, he sought to explain the elastic nature of the medium by a new constitutive hypothesis in which the particles of the medium were taken to be "very small bubbles, in which the subtle matter whirls in a circular motion."

The skeleton of this notion was Cartesian, but Euler went further by proposing that the elastic compressibility of the medium be interpreted as a measure of the deformability of these bubbles against centrifugal forces.[17] In spite of the superficial similarities, Euler can hardly be credited with any anticipation of the orbiting electrons of modern atomic structure; and since only the net elasticity of the medium and its mean density entered quantitatively in his calculations, the artificiality of his hypothesis did not make itself felt. Euler's 1727 version of Newton's formula for the speed of sound, which he stated baldly without proof, included an unexplained extra factor of $4/\pi$, which was just about as useful, and just as spurious, as Newton's data manipulations had been in bringing the theoretical prediction of the speed of sound into more comfortable agreement with experiment.

Nearly half a century later, Johann Heinrich Lambert (1728–77), who is better known for his real accomplishments in the fields of light and heat, and whose name is now attached to the standard unit of brightness, made one of the last attempts to resolve the Newtonian dilemma by artificially doctoring up the medium. According to Lambert, ordinary air consisted of heterogeneous parts, some of which "transmitted their weight but not their elasticity," and he claimed that these "impurities" ought to be excluded in reckoning the density of the "pure" air which alone served to transmit the sound.[18] The only way he could use this hypothesis, however, was to work backward from the measured sound speed to the outlandish conclusion that "pure" air is about a third less dense than ordinary air!

In 1762, after d'Alembert, Euler, and Lagrange had each made substantial contributions toward rounding out a definitive formulation of the basic equations of acoustics (see below), Lagrange made the first suggestion about the nature of the medium that was to contribute to the resolution of the Newtonian speed dilemma. Lagrange pointed out that sound could be expected to propagate in exactly the same way, but with altered speed, if the air were less easily compressible than Newton had assumed; that is, if the condensations were not linearly proportional to the pressure. Then, guided more by his intuitive feeling for mathematical relationships than by rationalization, Lagrange suggested that it would be "the most natural case" for pressure to vary as some power of the density. Moreover, it was easy for him to deduce that good agreement with experiment would prevail if the pressure varied as the 4/3 power of the density but, since he could offer no plausible physical reason why such a power law should hold, he was content to abandon the hypothesis as a "fleeting conjecture."[19]

The Evolution of the Adiabatic Hypothesis

The thermal modifications of the "gas law" that were needed to back up Lagrange's "fleeting conjecture" took a long time to mature. They finally did, but only after a tortuous development in which they were linked with first one and then another of the major heat problems that were painfully worked out during the late eighteenth and early nineteenth centuries. Even the simple pressure-density relation on which Newton had relied, now commonly but not very justifiably referred to as "Boyle's Law" [pv = constant], was ambiguous in origin. Richard Towneley (1628–1706/7)[20] was the first to point out to Boyle that the data given in his essay of 1660 on "the spring of the air" clearly indicated "the pressures and expansions to be in reciprocal proportion." Boyle acknowledged Towneley's contribution and the confirming advice of two others in his *Defence* annexed to the reprint edition (1662) of his famous tract.[21] A similar study of the compressibility of gases was made later and independently in France by Edme Mariotte (1620–84), and his exposition of the subject had a clarity that Boyle's had lacked.[22] Partly on this account, the "Towneley-Boyle" law was known on the Continent throughout most of the eighteenth and nineteenth centuries, and perhaps in some places even yet, as the "loi de Mariotte."

The gas thermometer held the secret of one of the needed modifications of the Towneley-Boyle-Mariotte gas law but it took a century and a half for its relevance to the acoustical problem to be appreciated. The thermal expansion of air had been exploited by Heron, Philo, and others for the activation of toys and automata, but the first *instrument* that made use of the phenomenon for the measurement of temperature was Galileo's air *barothermoscope* (ca. 1592). This device was modified and improved by some of Galileo's followers, including Ferdinand II, and it was used in many of the heat experiments conducted under the auspices of the short-lived Accademia del Cimento (see pt. 2, n. 68).

By the end of the seventeenth century, in the hands of the French physicist and student of friction, Guillaume Amontons (1663–1705), Galileo's crude instrument had become a constant-volume gas thermometer; and Amontons was led to the important conclusion "that unequal masses of air increase equally in pressure with equal degrees of heat, and contrariwise."[23] There were two gems in this simple statement, but Amontons seems not to have recognized one nor to have given the other enough polish. In the light of Boyle's discussions of the "spring of the air" and its relation to pressure, and of Newton's theoretical analysis of the speed of sound, Amontons might very well have immediately drawn the conclusion that the speed of sound ought to increase with temperature. This inference was entirely missed, however, not only by Amontons but by all the other scientists of the eighteenth century; and even the experimental discovery (ca. 1740) of the gross effects of temperature on the speed of sound, through the experiments of Bianconi and Condamine (see pt. 2, nn. 77, 78), did not bring it forth.

Amontons's unpolished jewel was the notion of absolute temperature, a concept that was doomed to be dallied with inconclusively for a century and a half before Kelvin formulated it definitively in 1848.[24] Boyle, Hooke, and Huyghens had each suggested independently (ca. 1665) that temperature might be scaled in terms of the proportional expansion of some working substance measured from a single fixed reference temperature. Amontons adopted part of this notion and chose the boiling point of water as his reference temperature; but he recorded the pressure changes he observed in arbitrary units rather than as fractional parts of the total pressure at the reference temperature. As a result, the closest he could come to the concept of absolute temperature was to observe that "the extreme cold of this thermometer is that which would reduce the air by its spring to sustain no load at all."[25]

The next step toward a temperature-modified gas law, and toward the absolute-temperature concept, was taken by Lambert three-quarters of a century later, when he used a constant-pressure form of the gas thermometer to parallel some of Amontons's earlier work. In the meantime, of course, there had been much interest and some progress in the development of a technology of thermometry. Lambert identified no less than nineteen different temperature "scales" that had already been proposed by the authors of different schemes of thermometry. Water, oils, various "spirits," and mercury, as well as air, were used for the working substance; and there were at least two centessimal scales based on use of the boiling and freezing points of water as fixed points, as had been suggested by Renaldini (1615-98) in 1694. One of these was the inverted scale, with 0° at the steam point and 100° at the ice point, proposed in 1742 by Andreas Celsius (1701-44). The right-side-up revision of this scale, now designated simply "Celsius," was suggested by Jean Pierre Christian (1683-1755) in 1743 and proposed more definitively in 1750 by a former colleague of Celsius, Marten Strömer (1707-70).[26] Of the absolute zero, Lambert said that "a degree of heat equal to zero is really what may be called absolute cold . . . [at which] the volume of the air is zero, or as good as zero."[27] Unfortunately, Lambert failed, as Amontons had, to state his results on the thermal expansion of air in such a way as to yield either a clear modification of the gas law or a firm basis for an absolute temperature scale.

Much the same can be said of the work (ca. 1787) of Jacques Alexandre César Charles (1746–1823), whose results were rendered indecisive by the fact that neither pressure nor volume were separately controlled in his experiments. Charles did not publish his results, but they came accidentally to the attention of Joseph Louis Gay-Lussac (1778–1850), to whom Charles owes the overly generous credit commonly assigned to him in this connection. In the account of his own work, Gay-Lussac did give an accurate statement of the linear law of thermal expansion for gases when he concluded "that, in general, all gases by equal degrees of heat, under the same conditions, expand proportionately just alike."[28] So little later as to indicate that it was a parallel effort, John Dalton (1766–1844) concluded similarly that "all elastic fluids under the same pressure expand equally by heat."[29]

The labored trail from Amontons to Gay-Lussac and Dalton stretched from the beginning to the end of the eighteenth century, and the prize at its end was merely the fact that the constant in the Towneley-Boyle law is proportional to temperature. Another equally important facet of the truth about the physics of gases—the adiabatic heating and cooling of a gas when it is suddenly compressed or expanded—had to be exposed and polished before it could serve either to resolve the dilemma about sound speed or play its part in the conquest of the caloric theory of heat conservation.

Anyone who uses a hand pump to inflate a football or a bicycle tire quickly learns that at least part of the work of compression appears as heat in the gas. The association between such thermal phenomena and the conservation of energy is so close, in fact, that adiabatic heating is now commonly presented in elementary textbooks as an axiomatic demonstration of the validity of the principle of energy conservation. There is ample precedent for this association, since each of the seven men who independently "discovered" the law of the conservation of energy during the period 1824 to 1845 turned immediately for support to the experiments on adiabatic heating and cooling.[30] Ironically enough, these same experimental facts had been used with equal satisfaction by the caloric theorists during the preceding half-century in support of their theory of heat conservation.[31] The caloric theory reached its climacteric with Carnot's classic paper[32] of 1824, but it moved rapidly toward eclipse during the following decades as the evidence mounted in support of the principle of energy conservation. The defendants of the caloric theory displayed admirable, if misguided, ingenuity in twisting and adapting it to explain the results of a mounting variety of new experiments. A late example of this kind of virtuosity was exhibited by Poisson, who showed in 1823 that one could derive a proper value for the speed of sound and some other valid relations concerning adiabatic behavior as a consequence of the single "caloric" hypothesis that heat content is a state variable depending only on pressure and density.[33]

As for the raw facts about adiabatic heating and cooling, these were exploited, observed, misexplained, inadequately measured, and generally misunderstood, in about that order, before enough understanding came to allow them to be used in a final resolution of the sound-speed dilemma. Exploitation came first, taking the form of a *fire syringe* said to have been used by the Dyaks of Borneo during the eighteenth century or earlier. With this device a small volume of confined air could be suddenly compressed and sufficiently elevated in temperature to ignite tinder. The origins of this remarkable bit of primitive technology remain obscure, but it became a popular lecture demonstration during the nineteenth century. Its late appearance in the European technical literature, however, makes it clear that news of it came to the West too late to influence the growth there of the adiabatic concept.

The first deliberate experimental observation of these effects was recorded by the Scottish physician William Cullen (1712–90), who noticed in the course of his work on evaporation that "A thermometer hung in the receiver of an air-pump, sinks always two or three degrees upon the air's being exhausted. After a little time, the thermometer in vacuo returns to the temperature of the air in the chamber, and upon letting air again into the receiver, the thermometer always rises two or three degrees above the temperature of the external air."[34] He did not try to explain these "curious *phaenomena,*" saying merely that "the experiments must be often repeated, and much diversified, before I can give the Society a proper account of them." The experiments were indeed repeated often during the next half-century, but with remarkably little diversification except in the explanations brought forward to explain the effects.

Only a year later Johann Christian Arnold (1724-65), at the University of Erlangen, devoted himself to a more extensive investigation of these effects. Arnold insisted that the cooling observed when the chamber was evacuated was a consequence of evaporation, in spite of the fact that he took pains to dessicate the receiver in some of his tests, and he attributed the complementary heating to the friction between the thermometer and the air entering the receiver.[35]

These effects lay fallow for the next two decades but there were other exciting developments on the theoretical and application frontiers of heat. Joseph Black (1728-99) sharpened up the distinction between *temperature* and the *quantity of heat* and introduced the concepts of specific heat and latent heat, James Watt (1736-1819) invented and patented his condensing steam engine, and the caloric theory was on the march. In his important publication of 1777 on *Pyrometrie,* Lambert took due notice of the Cullen and Arnold experiments, but he did not bother to repeat them, saying that they "had nothing which would have astonished me very much." According to Lambert, the existing "fire particles" were pumped out of the receiver with the air, but the reduced density of the fire particles left in the receiver was quickly compensated by the fire particles "pressed out of the glass [of the receiver] and the plate [of the air pump] into the thinned air in order to replace the outflow."[36a] Both the initial fall in temperature and its subsequent equilibration were thus glibly explained.

Arnold had been fully aware of the transient character of the phenomena he observed, and he had shown that the temperature changes were much smaller when the air was admitted to the receiver or withdrawn from it slowly. Lambert extended his remarks on this topic, pointing out that these "momentary effects can be considerable in their results." In doing so he had in his hands for a moment the phenomenological aspect of the experiment that is central to the acoustical problem—but, alas, he failed to see its acoustical relevance. When "the air is compressed momentarily into a doubly smaller space," Lambert pointed out, "its density, as well as its heat, is doubled, and so its elasticity must not double but must be four times greater. It retains this fourfold size not long, however, . . . ever so quickly the fourfold elasticity comes down again to the two-fold."[36b] It was just this extra increase of elasticity, of course, that was needed to explain the excess of sound speed over the value that Newton had calculated—an increment of elasticity stemming from the minute temperature changes induced by the compressions and rarefactions that constitute the sound wave. Lambert's remarks could, therefore, have provided just as firm a basis for attacking the sound-speed problem as the similar suggestion which LaPlace made to Biot twenty-three years later; except that LaPlace *did* appreciate the acoustical relevance of the temperature-induced increment of elasticity, whereas Lambert missed it.

In the meantime, the Cullen-Arnold thermometer-in-a-bell-jar experiment came in for more of the same kind of repeated attention that had been devoted to the bell-in-a-vacuum experiment a century earlier. Horace Benedict de Saussure (1740–99) in his influential *Hygrométrie*[37] of 1783, and Marc Auguste Pictet (1752–1825) in a now rare essay *Sur le feu*[38] of 1790, gave new accounts of similar experiments, and each adopted Lambert's explanation of the results. Erasmus

Darwin (1731–1802), grandfather of the great naturalist Charles Darwin, repeated the Cullen-Arnold experiments "twelve to fourteen" years before he published in 1788 one of the more articulate reviews of the experimental status of this topic.[39] Darwin's explanation of the usual effects differed from Lambert's, but the difference did more to show the nimble versatility of the caloric theory than to deepen understanding. More originality was displayed by Darwin's perceptive suggestion that the prevailing low temperature at high elevations was a consequence of the expansion cooling of air carried upward by wind circulation. The roster of these early adiabaticists was rounded out by Ludwig Achim von Arnim (1781–1831), who discussed once more the classic experiments,[40] and by John Dalton (1766–1844), who brought this episode to a close with his definitive review[41] of 1802.

NOTES

Introduction

1 Sir Francis Bacon (Baron of Verulam, Viscount St. Alban), *Sylva Sylvarum* [pub. posth., 1627], Century 2, §114, p. 390, in vol. 2 of *The Works of Francis Bacon*, ed. J. Spedding, R. L. Ellis, and D. D. Heath (London: Longman and Co., et al., 1857).

2 Ernst Mach, *Die Mechanik* (Leipzig, 1883, 1889, etc.), trans. T. J. McCormack (London and LaSalle, Ill.: Open Court Pub. Co., 1893, . . . 5th English ed., 1942).

1. Origins in Observation

1 See, for example, Boethius, *De Institutione Musica*, Book 1, §10, p. 196, l. 18 to p. 197, l. 3 of the text edited by G. Friedlein (Leipzig: B. G. Teubner, 1867); trans. I. E. Drabkin, in M. R. Cohen and I. E. Drabkin, *A Source Book in Greek Science* (New York: McGraw-Hill Co., 1948), p. 298. Abbreviated hereafter as *SBGS*.

2 Hermann Diels, *Die Fragmente der Vorsokratiker*, 1st ed., 1903; 5th ed., ed. W. Kranz (Berlin: Wiedmann, 1934–37); a complete translation of these fragments is available in Kathleen Freeman, *Ancilla to the Pre-Socratic Philosophers* (Cambridge, Mass.: Harvard University Press, 1948). (a) Heraclitus, frags. 101a and 107 (Freeman, pp. 31–32). (b) The bracketed clause is translated differently by Freeman and by A. Fairbanks in *The First Philosophers of Greece* (London, 1898), p. 25. (c) Anaxagoras, frag. 21 (p. 86). (d) Philolaus, frag. 4 (p. 74). (e) Archytas, frag. 1 (p. 78).

3 Plato, *Republic*, Book 7, pp. 525A–530D (Steph. 2) [the marginal pagination of almost all Platonic texts is based on that of the Greek and Latin edition printed by Henri Estienne (Stephanus) at Geneva in 1578]; in vol. 3 of *Dialogues of Plato*,

translated by Benjamin Jowett (New York and London: Oxford
University Press, 3d ed., 1892).

4   Aristotle, *De Caelo*, Book II.9; p. 290^b, lines 16-32; translated by
J. L. Stocks in vol. 2 (1930) of the Oxford Aristotle, a distinguished series of translations edited by W. D. Ross and published
by the Oxford University (Clarendon) Press.

5   Alexander of Aphrodisias, *Commentary on Aristotle's Metaphysics*, A5, p. 542^a 5-18; translated in Sir Thomas L. Heath,
Greek Astronomy (London: J. M. Dent & Sons, 1932), p. 34.

6   A typical account of the hammer legend, and of the invention of
the monochord, is given by Boethius, *De Institutione Musica*,
Book I, §§ 10-11; pp. 197-98 of the Friedlein text (see n. 1,
above).

7   Theon of Smyrna, *On Mathematical Matters Useful in Reading
Plato*, pt. 2, § 13; p. 100, ll. 7-11 of the text edited and trans-
lated into French by Jean Dupuis (Paris: Librairie Hachette et
Cie, 1892).

8   Archytas, frag. 1; quotations from (p. 79) the Freeman translation of Diels's *Fragmente* (see n. 2, above).

9   Aristotle, *De Sensu*, chap. 7, pp. 448^a 20-22; translated by J. I.
Beare as part of the *Parva Naturalia*, vol. 3 of the Oxford Aristotle
(Oxford: Clarendon Press, 1908).

10 Theophrastus of Eresus, *Fragment 89*: 11, p. 439, ll. 19-21; in
Theophrastus, *Opera Omnia*, Greek text revised, with parallel
translation into Latin, by Friedrich Wimmer (Paris: A. F. Didot,
1866). This fragment is quoted in its entirety by Porphyry in his
Commentary on Ptolemy's Harmonics, the Greek text of which
has recently been edited by Ingemar Düring (Göteborg: Elanders,
1932; and *Göteborgs Högskolas Årsskrift* 38, pt. 2, pp. xliii, 217
(1932). I am indebted to William Chase Greene for assistance with
the translation of this passage and its context.

11 Jean Henri Hassenfratz, "Mémoire sur la propagation du son,"
Annales de Chimie, I$^\circ$ ser., vol. 53, pp. 61-71 (1805). An incorrect
reference to Biot in this connection has propagated itself in several historical accounts of nineteenth-century acoustics. J. B. Biot
did carry out similar and more extensive experiments, which he
described in a paper read before the Institut National in November 1808 and published in the *Mémoires de Physique et de Chimie
de la Société d'Arcueil* (Paris: Bernard, 1809), 2:405-23, and also
in vol. 2, pp. 7-8, of his *Traité de Physique* (Paris: Deterville,
1816).

12 Theon of Smyrna, *On Mathematical Matters* . . . , pt. 2, § 12; p.
92, ll. 19-21 of the Dupuis text and translation (see n. 7, above).

13 Philolaus, frag. 6; quotation (p. 74) from the Freeman translation of Diels's *Fragmente* (see n. 2, above).

14 Aristotle, *Problemata* 19.41; p. 921^b 1-13; trans. E. S. Forster in vol. 7 of the Oxford Aristotle (Oxford: Clarendon Press, 1927).

15 Euclid, *Sectio Canonis*, introduction; trans. I. E. Drabkin in *SBGS*, p. 291 (see n. 1, above).

16 See, for example, Proclus Lycius (surnamed Diadochus), *Commentary on Euclid's Elements*, trans. Thomas Taylor (London, 1792); a selection also translated (from the Friedlein edition) by I. E. Drabkin in *SBGS*, p. 37 (see n. 1, above). This commentary is chiefly notable as a history of pre-Euclidean geometry, but it is also the chief surviving source of information about Euclid's life.

17 Plato, *Apology*, p. 31A (Steph. 1); trans. H. N. Fowler in #36, Loeb Classical Library (Cambridge, Harvard University Press; and London: Heinemann, 1914, and nine reprintings to 1947).

18 Plato, *Republic* 7.531B (Steph. 2), trans. B. Jowett (see n. 3, above).

19 Plato, *Timaeus*, p. 67B (Steph. 3); in vol. 3 of *Dialogues of Plato*, trans. B. Jowett (see n. 3, above).

20 Theophrastus, *On the Senses;* trans. G. M. Stratton in *Theophrastus and the Greek Physiological Psychology* (London: Allen & Unwin; New York: Macmillan, 1917). (a) Alcmaeon, § 25 (Stratton, p. 89); (b) Empedocles, §9 (Stratton, p. 73); (c) Anaxagoras, §§ 28-30 (Stratton, pp. 91-92).

21 Lucretius, *De Rerum Natura*, Book 4, ll. 524-26; trans. W. H. D. Rouse in #181, Loeb Classical Library (Cambridge, Harvard University Press; and London, Heinemann, 1924; 3d ed. revised and reprinted, 1947).

22 Aristotle, *De Anima*, Book 2.8; p. 420^a 4-5; trans. J. A. Smith in vol. 3 of the Oxford Aristotle (Oxford: Clarendon Press, 1931).

23 C. B. Boyer, "Aristotelian References to the Law of Reflection," *Isis* 36 (1946): 92-95.

24 Aristotle, *Problemata* (see n. 14, above), Book 11.23; p. 901^b 22.

25 Aristotle, *Meteorologica*, Book 2.9; p. 369^b 7-9; translated by E. W. Webster in vol. 3 of the Oxford Aristotle (Oxford: Clarendon Press, 1923).

26 Pliny the Elder, *Naturae Historiarum*, Book II.55, § 142; trans. H. Rackham in #330, Loeb Classical Library (Cambridge, Harvard University Press; and London, Heinemann, 1938; rev. ed., 1949).

27 Lucretius, *De Rerum Natura*, Book 6, ll. 164-72 of the translation into English verse by Charles Frederick Johnson (New York: DeWitt C. Lent & Co., 1872).

28 Aristotle, *De Anima*, Book 2.8; pp. 419^b 20-28; 420^b 29-33.

29 Democritus, frag. 145; p. 105 of the Freeman translation of Diels's *Fragmente*.

30 Diogenes Laertius, *Lives and Opinions of the Philosophers*, Book VII, chap. 1, § 158; trans. R. D. Hicks, in 2 vols., #184 (Bks. 1-5) and #185 (Bks. 6-10), Loeb Classical Library (Cambridge, Harvard University Press; and London, Heinemann Ltd., 1925; revised and reprinted, 1950). The passage quoted is an improved translation kindly supplied by Professor C. A. Truesdell.

31 Vitruvius, *De Architectura*, Book 5, chap. 3, §§ 6-7; translation (into German) in Ernst Mach, *Die Mechanik* (see Intro., n. 2, above). This quotation is taken from McCormack's English translation, which seems to afford a smoother technical rendition of the passage than do the more recent Morgan (1914) and Granger translations from the primary MSS. Cf. below, note 52.

32 Seneca, *Naturalium Quaestionum*, Book 2.6, §§ 3-4; trans. John Clarke (London, Macmillan and Co., 1910).

33 Democritus; quoted in translation from a source, presumably Posidonius, in Milton C. Nahm, *Selections from Early Greek Philosophy* (New York: Appleton-Century-Crofts, 1947), p. 189.

34 Aristotle, *Problemata* (see n. 14, above), Book 11.23, p. 901^b 16-17; and Book 11.51, p. 904^b 27-28.

35 Lucretius, *De Rerum Natura*, Book 4, ll. 549-56.

36 Theophrastus, *On the Senses*, §§ 51-53 (Stratton, pp. 111-13).

37 Epicurus, in Diogenes Laertius, *Lives and Opinions*, Book 10, § 53.

38 [?Straton of Lampsacus?], *De Audibilibus*, p. 800^a 1-14; translated (from the Bekker text) by W. S. Hett in *Aristotle—Minor Works*, #307 of the Loeb Classical Library (Cambridge: Harvard University Press; and London, Heinemann, 1936); also translated (from the Prantl text) by T. Loveday and E. S. Forster in vol. 6 of the Oxford Aristotle (Oxford: Clarendon Press, 1913). The first portion of the passage quoted is taken from the Hett translation, the remainder from the Oxford version.

39 Aristotle: compare the "not audible" passage in *Problemata*, Book 11.58, p. 905^b 4 and 10; with the "hear also" passage in *De Anima*, Book 2.8, p. 420^a 12; and with the discussion of hearing in fishes in *Historia Animalium*, Book 4.8, pp. 533^b 1-534^a 12, translated by D. W. Thompson in vol. 4 of the Oxford Aristotle (Oxford: Clarendon Press, 1910).

40 Georges Rodier, *La Physique de Straton de Lampsaque*, pp. 47-50 (Paris: Germer Baillière et Cie, 1890).

41 Theophrastus, *On the Senses* (see n. 20, above): further reference to Democritus in §55 (Stratton, p. 115).

42 Boethius, *De Institutione Musica*, Book 1; §3, p. 189, ll. 22-23; and §14, p. 200, ll. 7-9 and 17-26 of the Friedlein text (see n. 1, above). Translated by I. E. Drabkin in *SBGS*, pp. 292-94.

43 Aristoxenus, *Elements of Harmonics*, Book 2; §33, ll. 3-6 and §44; trans. Henry S. Macran, pp. 189 and 197-98 (Oxford: Clarendon Press, 1902).

44 Leon Boutroux, "Sur L'Harmonique Aristoxenienne," *Revue Générale des Sciences* 30 (1919): 265-74.

45 Ptolemy, *Harmonics*, Book 2, chap. 13; Greek text newly edited by Ingemar Düring (Göteborg: Elanders, 1930); German translation by Düring in *Ptolemaios und Porphyrios über die Musik* (Göteborg: Elanders, 1934), and in *Göteborgs Högskolas Årsskrift* 40, pt. 1 (1934): 1-293.

46 Gioseffo Zarlino, Le Istitutione Armoniche, Sopplimenti Musicali, Book 4 (Venice: Francesco de' Franceschi, 1588).

47 Charles Burney, *A General History of Music from the Earliest Ages to the Present* [1789] *Period*, 4 vols. (London, 1776-89; 2d ed. of vol. 1, 1798; republished in 2 vols., with critical and historical notes by Frank Mercer, London: G. T. Foulis & Co., 1935). The references to Ptolemy occur on pp. 356ff. of vol. 1 of the Mercer edition.

48 Bartolomeo Ramos de Pareja, *Musica Practica* (Bononiae [Bologna], 1482; republished by Johannes Wolf, in *Publikationen der Internationalen Musikgesellschaft, Beihefte*. II. (Leipzig: Breitkopf & Härtel, 1901).

49 For a very readable account of these problems of musical scales and temperament, see Wilmer T. Bartholomew, *Acoustics of Music*, chap. 4, pp. 163-98 (New York: Prentice-Hall, 1945).

50 Aristotle, *Problemata* (see n. 14, above): (a) 11.25, 901^b 30-34; (b) 11.7, 899^b 18-22; (c) 11.52, 904^b 33-36; (d) 11.49, 904^b 15-22; and 11.58, 905^a 35-905^b 2.

51 Lucretius, *De Rerum Natura*, Book 4, ll. 560-614.

52 Vitruvius, *De Architectura*, Book 5, chap. 8. This important passage affords a capital illustration of the semantic hazards of technical translation. There is obviously a difference in the suggestiveness of such word choices as sound *driven back* or *reflected*, of the voice *overwhelming* a following utterance or *interfering* with it, and so on. Such distinctions are as likely, and sometimes as soundly, based on the gleam in the translator's eye as on differences in the Latin texts. The reading quoted here is my

rendition based on a comparison of the earlier English translations
with the scholarly translation by Frank Granger of the Harleian
MS 2767, in 2 vols. (#251 and #280) of the Loeb Classical Library
(Cambridge: Harvard University Press; and London, Heinemann,
1931–33; vol. 1 reprinted, 1945).

53   Joshua 6:20.

54   Herodotus, Book 4, §200; translated by A. D. Godley in vol. 2
     (of 4), #118 of the Loeb Classical Library (London: Heinemann;
     and New York: G. P. Putnam's Sons, 1921).

55   Vitruvius, *De Architectura* (see n. 52, above): (a) Book 10, chap.
     16, §§9–10; (b) Book 1, chap. 1, §8, and Book 10, chap. 12, §2.
     Granger translation.

56   Heron of Alexander, *A Treatise on Pneumatics*, §50; p. 72 of
     the translation (with illustrations) by Jos. G. Greenwood (Lon-
     don: Taylor, Walton and Maberly, 1851).

57   Heron of Alexander, *Catoptrics*, §4; p. 324, l. 16 to p. 328, l. 8 of
     the Latin text, edited and translated into German by Wilhelm
     Schmidt (Leipzig: B. G. Teubner, 1900): the "shortest-path"
     proof is also given in the portion translated by I. E. Drabkin in
     SBGS, p. 264.

58   Cleomedes, *De Motu Circulari Corporum Caelestium* 1.10, §52;
     pp. 90.20–91.2 and 94.24–100.23 of the Ziegler edition. Trans-
     lated in Heath's *Greek Astronomy* (see n. 5, above), pp. 109–12.
     For general material on Hipparchus, Eratosthenes, et al., see Thos.
     L. Heath, *History of Greek Mathematics* (Oxford: Clarendon
     Press, 1921); or *Greek Mathematical Works* (trans. Ivor Thomas)
     in the Loeb Classical Library.

59   Samuel Eliot Morison, *Admiral of the Ocean Sea; A Life of
     Christopher Columbus*, vol. 1 (of 2), pp. 45–47, 56–58, 86–87.
     (Boston: Little, Brown and Co., 1942); see also D. Merejkowski,
     Romance of Leonardo da Vinci, trans. B. G. Gurney (New York,
     Random House, 1928), p. 370.

60   A good account of this legendary incident is given by Vitruvius
     in *De Architectura* (see n. 52, above), Book 9, preface, §§9–12.

61   Roger Bacon, *Opus Majus*, pt. 4, First Distinction, chap. 3, at pp.
     123–24 of the translation by Robert B. Burke (Philadelphia: Uni-
     versity of Pennsylvania Press, 1928).

62   Aristoxenus, *Elements of Harmonics* (see n. 43, above), II. §33,
     ll. 1–2; p. 189 of the Macran translation.

2. Origins in Experiment

1   Guy Le Strange, *Baghdad during the 'Abbasid Caliphate* (Oxford:

Clarendon Press, 1900; reprinted 1924). The style adopted here for transliteration of proper-name references for the medieval period follows George Sarton, *Introduction to the History of Science*, 3 vols. in 5 (Baltimore: Williams & Wilkins, vol. 1, 1927; vol. 2 in 2, 1931; vol. 3, part 1, 1947, part 2, 1948). Hereafter cited as Sarton, *Intro. Hist. Sci.*

2   Paul Lacroix, *Science and Literature in the Middle Ages* (London: Bickers & Son, 1878), p. 108.

3   The method of reckoning dates from the Incarnation was not introduced until the sixth nor widely adopted until the tenth century. Modern studies of chronology indicate that the proposer of this reference date, Dionysius Exiguus (fl. 525–d. ca. 540), made a mistake in his computations. Correcting the error leads to the paradox that Christ was born about 8 B.C. and died A.D. 28. Cf. Sarton, *Intro. Hist. Sci.*, 1:236 and 429. See also Charles Eugène Cavaignac, *Chronologie de l'Histoire Mondiale*, pp. 14, and 197–211 (Paris: Payot, 1925; 2d ed., 1934); and W. M. Calder, "The Date of the Nativity," *Discovery* 1 (London, 1920): 100–03.

4   Sarton, *Intro. Hist. Sci.*, 1:21–29.

5   E. G. P. Wyatt, *St. Gregory and the Gregorian Music* (London: published for the Plainsong and Medieval Music Society, 1904); also Amédée Gastoué, "L'art grégorien," in *Les Maitres de la Musique* (Paris: Félix Alcan, 1911).

6   A good bit of prime material dealing with music in the Middle Ages has by now become available in modern-language translations, but a canvass of the original sources in pointed search for clues to the understanding of the *physics* of sound phenomena in general is still wanting. Invaluable collateral material is to be found in the translations and commentaries of the Irish orientalist Henry George Farmer, and these have furnished the point of departure for much of the following discussion. See, for example, Farmer's *Sources of Arabian Music: An Annotated Bibliography* (Bearsden, Scotland: Privately printed, 1940), and his *History of Arabian Music* (London: Luzac & Co., 1929).

7   H. G. Farmer, *Historical Facts for the Arabian Musical Influence* (London: W. Reeves, 1930): (a) chap. 1; (b) appendix 41, pp. 317–20.

8   H. G. Farmer, "Greek Theorists of Music in Arabic Translation," *Isis* 13 (1930): 325–33.

9   H. G. Farmer, *Al-Fārābī's Arabic-Latin Writings on Music* (the second fascicule in *Collection of Oriental Writers on Music*, ed. H. G.

Farmer) (Glasgow: The Civic Press, 1934), pp. 37–41; discussion of authorship also in *Journal of the Royal Asiatic Society* (1933), pp. 307–11.

10 Al-Fārābī (or Thābit ibn Qurra?), *De Ortu Scientiarum.* This passage is translated (with the Latin text exhibited in a footnote) by Lynn Thorndike, *History of Magic and Experimental Science* (New York: Macmillan, 1923), 2:81.

11 Al-Fārābī, Kitab al-Mūsīqī al-Kabīr (Grand traité de la musique), French translation by Baron Rodolphe d'Erlanger, in *La Musique Arabe*, 5 vols. (Paris: Librairie Orientaliste Paul Geuthner, 1930, 1935, 1938, 1939, 1949): (a) 1:81; (b) 1:82–83; (c) 1:83–85; (d) 1:262–76; (e) 2:26–52; (f) 1:86–93.

12 H. G. Farmer, "Clues for the Arabian Influence on European Musical Theory," *Journal of the Royal Asiatic Society* (1925), pp. 61–80.

13 Many primary references to the copious literature on this subject are cited in Jules Combarieu, *Histoire de la Musique* (Paris: Armand Colin, 1913, 8th reprinting, 1948), 1:153–61, 233–41; for example, Plato *Laws* 2.652–74, 7.798, 800, 802, 812; *Republic* 3.398–403, 410–11; or Aristotle *Politics* 8.5-7 (1339^a–42^b).

14 Ignatius de Mouradja d'Ohsson, *Tableau Général de l'Empire Othoman* (Paris: De l'Imprimerie de Monsieur, 1791), vol. 4, pt. 2, pp. 280–81.

15 Duncan B. MacDonald, "Emotional Religion in Islām as affected by Music and Singing. Being a Translation of a Book of the *Iḥyā 'Ulum ad-Din* of al-Ghazzālī with Analysis, Annotation, and Appendices," *Journal of the Royal Asiatic Society* (1901), pp. 195–252, 705–48, and (1902), pp. 1–28; (a) (1902), p. 13; (b) (1901), p. 199; (c) p. 201; (d) p. 208; (e) p. 218.

16 From the *Kitab al-'iqd al-farid* of Ibn 'Abd Rabbihi (A.D. 860–940) translated by H. G. Farmer in *Music: The Priceless Jewel*, (Bearsden, Scotland: Published by the author, 1942), p. 6; also in *Journal of the Royal Asiatic Society* (1941), p. 27.

17 Abu'l-Faraj al-Isfahānī, *Kitāb al-aghānī al-kabīr* [The grand book of songs], Sāsī ed. of the text, 21 vols. (Cairo, 1905–06); parts in French translation, E. M. Quatremère, *Journal Asiatique* [2°] 16 (1835): 385–419, 497–545.

18 Al-Mas' ūdī, *Muruj al-dhahab* [Meadows of gold], text ed. and trans. C. Barbier de Meynard and Pavet de Courteille, in *Les Prairies d'or*, 9 vols. (Paris, 1861–77).

19 A. Caussin de Perceval, "Notices anecdotiques sur les principaux musiciens arabes des trois premiers siècles de l'islamisme," *Journal Asiatique* [7°] 2 (Nov.-Dec. 1873): 397–592.

20 Ikhwān al-Safā [Brethren of Purity], *Risāla on music, according to Al-Majriti* [fl. in Andalusa, d. 1007], trans. Friedrich Heinrich Dieterici, in *Die Propaedeutik der Araber im Zehnten Jahrhundert* (Berlin: E. S. Mittler und Sohn, 1865), pp. 100-53: (a) p. 104; (b) p. 105; (c) p. 104; (d) p. 105; (e) p. 109; (f) p. 106.

21 P. Casanova, "L'incendie de la bibliothèque d'Alexandrie par les Arabes," *Comptes Rendus des Séances de l'Académie des Inscriptions et Belles-Lettres* (Paris, 1923), pp. 163-71.

22 Cf. Sarton, *Intro. Hist. Sci.*, 1:707-09. Al-Bīrūnī's translators seem to have been more attracted to other phases of his work, so I have not been able to verify this reference to his observations on sound.

23 Ibn Sīnā (Avicenna), *Kitab al-Shifā*, chap. 12, discourse 1, art. 4, on p. 123 of Baron Rodolphe d'Erlanger's translation (Fr.), in *La Musique Arabe*, vol. 2 (Paris: Librairie Orientaliste Paul Geuthner, 1935).

24 H. G. Farmer, in his *Historical Facts* (see n. 7, above), pp. 72-82, argues authoritatively the inadequacy of the evidence to resolve this question.

25 The words of the hymn are:

> UTqueant laxis REsonare fibris
> MIra gestorum FAmuli tuorum,
> SOLve polluti LAbii reatum,
> Sancte Joannes.

26 Thomas Francis Carter, *Invention of Printing in China and Its Spread Westward* (New York: Columbia University Press, 1925; rev. ed., 1931), pp. 159-79; reviewed in *Isis* 8 (1925): 371.

27 Theophilus, called also Rugerus, *An Essay Upon Various Arts, in Three Books*, Latin text edited, translated, and annotated by Robert Hendrie (London: John Murray, 1847): (a) Book 3, chap. 86, p. 371; (b) Book 3, chap. 85, p. 363.

28 Adelard of Bath, *Quaestiones Naturales*, chaps. 21 and 22, English translation by Hermann Gollancz, in *Dodi Ve-Nechdi* [Uncle and Nephew] of Berachya, pp. 112-14 (London: Oxford University Press, Humphrey Milford, 1920). C. H. Haskins (see following note) refers to this as "a careless English version," but the quoted passages conform agreeably with the paraphrase given by Thorndike, *History of Magic* (see n. 10, above), 2:32. A sound Latin text is edited by M. Müller in *Beiträge zur Geschichte der Philosophie und Theologie des Mittelalters* Band 31, Heft 2 (Münster: Aschendorffschen Verlagsbuchhandlung, 1934).

29 There is a copious literature dealing with the mediaeval trans-

lators. See, for example, C. H. Haskins, *Studies in the History of
Mediaeval Science* (Cambridge: Harvard University Press, 1924);
or, more briefly, A. C. Crombie, *Robert Grosseteste and the
Origins of Experimental Science* (Oxford: Clarendon Press, 1953),
pp. 35–43; and, of course, Sarton, *Intro. Hist. Sci.*

30   Averrois Cordubensis, *Commentarium Magnum In Aristotelis De
Anima Libros*, Book 2, §97, p. 278, ll. 70–74, of the definitive
Latin text edited by F. Stuart Crawford (Cambridge, Mass.:
Mediaeval Academy of America, 1953). I am obliged to Professor
Crawford for the English translation of various passages from this
commentary, including the one quoted here.

31   See *The Medieval Science of Weights (Scientia de Ponderibus)*,
edited with introductions, English translations, and notes by
Ernest A. Moody and Marshall Clagett (Madison: University of
Wisconsin Press, 1952); also Pierre Duhem, *Les Origines de la
Statique* (Paris: Librairie Scientifique A. Hermann, 1905), pp.
99–108 and 354–58; or, for more temperate judgments on the
contributions of Jordanus, see Moritz Cantor, *Vorlesungen über
Geschichte der Mathematik* (Leipzig: B. G. Teubner, 1892), 2:55,
or Ernst Mach, *Science of Mechanics*, 5th English ed. (La Salle,
Ill.: Open Court Pub. Co., 1942), pp. 97–103.

32   A. C. Crombie, *Robert Grosseteste* (see n. 29, above): (a) p. 114;
(b) p. 216, n. 6.

33   Robert Grosseteste, *De generatione sonorum*, p. 7 of the Latin
text in *Die Philosophischen Werke des Robert Grosseteste*, ed.
Ludwig Baur [Band 9 of *Beiträge z. Ges. d. Philos. d. Mittelalters*
(Münster: Aschendorffschen Verlagsbuchhandlung, 1912)]. I am
indebted to Dr. J. A. Bradshaw for the English translation of this
passage.

34   Roger Bacon, *Opus Tertium*, Latin text edited by J. S. Brewer, in
*The Chronicles and Memorials of Great Britain and Ireland during
the Middle Ages* (London: Longman, Green, Longman, and
Roberts, 1859), vol. 15. See also *The Opus Majus of Roger
Bacon*, English translation by Robert Ellis Burke, 2 vols. (Phila-
delphia: University of Pennsylvania Press, 1928).

35   For tabulated references to the *Epistola . . . de magnete*, see
Sarton, *Intro. Hist. Sci.*, vol. 2, pt. 2, p. 1030. At least two
English translations are available: Sylvanus P. Thompson, "Petrus
Peregrinus de Maricourt and his Epistola de magnete," *Proceed-
ings of the British Academy (London)* 2 (1905–06): 377–408, and
also beautifully printed in a separate limited edition; and H. D.
Harradon, "Some early contributions to the history of geomag-
netism—I," *Terrestrial Magnetism and Atmospheric Electricity* 48

(1943): 3–17. For a short synopsis, see Paul F. Mottelay, *Bibliographical History of Electricity and Magnetism* (London: Charles Griffin & Company, 1922), pp. 46–53.

36 J. Combarieu, *Histoire de la Musique* (see n. 13, above): (a) p. 274; (b) pp. 277–81.

37 H. L. F. Helmholtz, *Sensations of Tone*, 3d English ed. (London: Longman, Green & Co., 1895: reprinted New York: Dover Publications, 1954), p. 283.

38 Safī al-Dīn 'Abd al-Mu'min, *Risalāt al Sharafiyya* [The Sharafian treatise on musical proportion], French translation by Baron R. d'Erlanger, with an introductory critique by H. G. Farmer, vol. 3, pp. 1–182 of *La Musique Arabe* (see n. 23, above): (a) §6, p. 10; (b) §4, p. 8; (c) §8, p. 11; (d) §9, p. 11; (e) §7, p. 10–11.

39 H. G. Farmer, *Sources of Arabian Music* (see n. 6, above), p. 56.

40 Al-Jurjani (?), *Commentary on Safī al Dīn's "Kitab al-Adwār"* [Book of musical modes], French translation by Baron R. d'Erlanger, vol. 3, pp. 185–565 of *La Musique Arabe* (see n. 23, above): (a) pp. 212–13.

41 See Lynn Thorndike, *History of Magic and Experimental Science* (New York: Columbia University Press, 1934), 3:446.

42 Democritus, see chap. 1, n. 33.

43 Geoffrey Chaucer, *House of Fame*, ed. Walter W. Skeat (London: Oxford University Press, 1900), Book 2: (a) ll. 257–262; (b) ll. 301–07, 315–19.

44 *The Notebooks of Leonardo da Vinci*, arranged, rendered into English, and introduced by Edward MacCurdy, 2 vols. (New York: Reynal and Hitchcock, 1938; "definitive" 1-vol. ed., New York: George Braziller, 1955). In the following citations the first number refers to vol. 1 of the 1938 edition, the second to the 1955 edition. (a) p. 614/600; (b) p. 622/608; (c) p. 63/59; (d) p. 543/524; (e) p. 544/525; (f) p. 71/65; (g) p. 528/508; (h) p. 586/570; (i) p. 74/68; (j) p. 284/268; (k) p. 70/64.

45 This and the following quotation are taken from the B. G. Guerney translation of D. Merejkowski's *Romance of Leonardo da Vinci* (New York: Random House, 1928), p. 333. They seem to be lyrical translations of fragments resembling those given by MacCurdy (see preceding note) on pp. 407/386 and 284/267.

46 Sir Francis Bacon (Lord Verulam), *Sylva Sylvarum* (posth. 1627), Centuries 2 and 3, in vol. 2 of *The Works of Francis Bacon*, ed. J. Spedding, R. L. Ellis, and D. D. Heath (London: Longman and Co. et al., 1857).

47 For a brief critique of Stevin's work, see Ernst Mach, *The Science of Mechanics*, McCormack trans., pp. 32–46, 109–10 (see chap. 1,

n. 31, above); also, *Simon Stevin, Principal Works*, vol. 1, general introduction—Mechanics, ed. E. J. Dijksterhuis (Amsterdam: C. V. Swets & Zeitlinger, 1955).

48 This legend has frequently been attacked on the ground of inconsistency. Vivianni's eye- and ear-witness biography of Galileo (*Racconto istorico della vita di Galileo*, Florence, 1654; and printed in vol. 1 of A. Favoro, ed., *Le Opere di Galilei Galileo* [Edizione Nazionale, 20 vols. in 21, Florence, Tipografia di G. Barbera, 1890-1909]) has Galileo observing the motion of Possenti's great bronze lamp at the age of eighteen, whereas this lamp was not installed in the cathedral at Pisa until December 1587. On the other hand, Galileo *did* spend three years at Pisa *after* Possenti's lamp was in place and, as Favaro has pointed out, there *were* lesser lamps in the cathedral during Galileo's student days and these *were* commonly drawn to one side for ease of lighting and then allowed to swing free. There is, therefore, a strong basis of plausibility for the legend, and the inconsistency in Vivianni's report of the story as Galileo told it to him may reveal no more than a slip in Galileo's recollection of one minor feature of the incident.

49 Galileo Galilei, *Discorsi a due nuove scienze* [Dialogues concerning two new sciences], English translation by Henry Crew and Alfonso de Salvio (New York: The Macmillan Co., 1914; paperbound reissue, New York: Dover Publications, 1952): (a) p. 95; (b) pp. 95-108; (c) pp. 100-01, 99; (d) pp. 101-02.

50 Marin Mersenne, *Traité de l'harmonie universelle; où est contenu la musique theorique & pratique des anciens & modernes, avec les causes de ses effets*, 2 pts. in 1 vol., octavo (Paris: Guillaume Baudry, 1627)[LC]; *Les Preludes de l'Harmonie*, 224 pp. (Paris: Henry Guenon, 1634); *Questions Harmoniques*, 276 pp. (Paris: Jaques Villery, 1634); *Harmonicorum libri*, 184 pp., bound with *Harmonicorum instrumentorium libri IV*, 168 pp. (fol., Lutetiae Parisiorum, sumptibus Guillelmi Baudry, 1635)[LC]; these page sequences were preserved when new front matter and a *Liber novus praelusorius* were added to constitute the *editio aucta* issued as: *Harmonicorum libri XII in quibus agitur de Sonorum Natura, Causis, et Effectibus: de Consonantiis, Dissonantiis, Rationibus, Generibus, Modis Cantibus, Compositione, orbisque totius Harmonicis Instrumentis* (Lutetiae Parisorium sumptibus Guillelmi Baudry, 1648).[LC, MH]

51 Marin Mersenne, *Cogitata Physico-Mathematico*, 2 vols., often in 1, $4°$ (Parisiis: sumptibus Antonii Bertier, 1644).

52 Marin Mersenne, *Harmonie Universelle, contenant la theorie et la*

pratique de la musique, où il est traité de la Nature des Sons & des Mouvemens, des Consonances, des Dissonances, des Genres, des Modes, de la Composition, de la Voix, des Chants, & de toutes sortes d'Instrumens Harmoniques (19 books in six page sequences, identified for reference and indexing by annexed letters *A* to *F*: 228.A = 3 books "de la nature du son & des Mouvemens" + 180.B = 2 books "de la Voix & des Chants" + 442.C [with errors] = 6 books "des Consonances & de la Composition" + 412.D [with errors] = first 6 books "des Instrumens" + 79.E = 7th book "des instrumens de percussin" + 68.F = "De l'utilité de l'harmonie"; usually 2 tom. in 1 vol., folio [Paris: Sebastien Cramoisy, 1636]). There are said to have been two other editions published in Paris by Richard Charlemagne and by Pierre Ballard (1636-37), but the only exemplars (three) I have been able to examine have been the Cramoisy edition.

53 Jacques-Charles Brunet, *Manuel du libraire*, 5th ed. (Paris: Firmin Didot Frères, Fils, et Cie., 1862), 3:1662.

54 *Harmonie Universelle* (see n. 52, above): (a) p. 2, "Preface generale au lecteur"; (b) Bk. 1, p. 14.A; (c) Bk. 1, p. 38.A; (d) Prop. ix, p. 44.F; (e) Bk. 2, prop. xv, p. 136.A; (f) Bk. 1, prop. xix, pp. − -46.D (as printed, but should be 50-51); (g) Bk. 1, prop. xviii, p. 45.D (should be 49); (h) Bk. 3, prop. xxi, corollary ix, p. 220.A; (i) Bk. 3, prop. vii, p. 123.D; (j) Bk. 3, prop. v, p. 169.A; (k) Bk. 3, prop. xviii, cor. vii, p. 150.D; (l) Bk. 3, prop. xviii, cor. iii, p. 149.D; (m) Bk. 3, prop. xvii, cor. iiii, p. 145.D; (n) Bk. 3, prop. xii, p. 134; prop. xvii, pp. 140, 142.D; (o) Bk. 1, prop. xxvii, and cor. 1, pp. 56-58.A; (p) Bk. 3, prop. xxi, p. 214.A; (q) Bk. 3, prop. xxi, cors. vi, vii, pp. 218-19.A; (r) Bk. 3, prop. xxi, pp. 214-15.A; (s) Bk. 3, prop. xxi, p. 215.A, and prop. vi, I. Advertissement, p. 41.F.

55 Bacon, *Sylva Sylvarum* (see n. 46, above), Century 3, §209, p. 416.

56 In *Cogitata* (see n. 51, above): (a) *Phaenomena Ballistica*, prop. xxxix (misprinted as xxxv), pp. 138-40; (b) *Harmoniae*, Bk. 1, art. 4, prop. v, p. 275; (c) ibid., prop. iv, corollary, p. 274; of passing interest in this connection, see F. V. Hunt, "High Speed Counting of Auditory Stimuli," *Review of Scientific Instruments* 7 (1936): 437; (d) ibid., prop. iv, p. 273-74; (e) *Phaenomena Ballistica*, prop. xv, "Duodecimo," pp. 44 ff.

57 *Harmonicorum libri XII* (see n. 50, above): (a) Second dedication, p. 1; (b) Bk. 2, "De causis sonorum," props. viii and ix, p. 12; (c) ibid., prop. xviii, p. 14; (d) ibid., prop. xxi, pp. 14-15.

58 Giovanni Battista Riccioli, *Almagestum Novum*, vol. 1, pt. 1, Bk.

2, cap. xx, prop. 1, p. 85, and prop. xi, pp. 86–87, 1 vol. in 2 pts., fol. (Bononiae: ex typographia haeredis Victorij Benatij, 1651); An entertaining account of these experiments is given by Alexander Koyré, "An Experiment in Measurement," *Proceedings of the American Philosophical Society* 97 (1953): 222–37.

59   Alexander J. Ellis, *History of Musical Pitch* (London: W. Trounce, 1880); also in *Journal of the Society of Arts*, 5 March and 2 April, 1880, and 7 January 1881; abstracted in the English translation of Helmholtz, *Sensations of Tone* (see n. 37, above), appendix 20, "Additions by the Translator," sec. H, pp. 493–513.

60   Leonhard Euler, *Tentamen novae theoriae musicae ex certissimis harmoniae principiis dilucide expositae*, chap. 1, §§9–13, pp. 6–8 (Petropoli: Academiae Scientiarum, 1739).

61   Joshua Walker, "Some Experiments and Observations concerning Sounds," *Philosophical Transactions of the Royal Society* (London) 20 (1698): 433–38. Also, according to R. T. Gunther, *Early Science in Oxford*, 14 vols. (imprint varies: most from Oxford: Clarendon Press, 1920–45), 4:173, an earlier "Discourse concerning sounds & Ecchoes, drawn up by Mr. Walker, was by him communicated & read" to the Philosophical Societie of Oxford on 23 February 1685/6.

62   G. E. Kahl, "Beobachtung der Schallgeschwindigkeit durch Coincidenzbeobachtungen," *(Schlömilch) Zeitschrift für Mathematik und Physik* 9 (1864): 65–69.

63   J. Bosscha, "Ueber ein Mittel die Schallgeschwindigkeit in einem eingeschlossenen Raume geraden zu messen" (Poggendorff) *Annalen der Physik* 92 (1854): 485–94.

64   K. R. Koenig, "Appareil pour la mesure de la vitesse du son," *Comptes Rendues* (Paris) 55 (1862): 603–05.

65   G. W. Pierce, "Piezoelectric Crystal Oscillators applied to the Precision Measurement of the Velocity of Sound in Air and CO_2 at HIgh Frequencies," *Proceedings of the American Academy of Arts and Sciences (Boston)* 60 (1925): 271–302.

66   J. M. A. Lenihan, "Mersenne and Gassendi," *Acustica* 1 (1951): 96–97.

67   P. Gassendi, *Opera Omnia* (Lyons, 1658), vol. 1, sec. 1, Bk. 6, p. 418. The passage quoted is translated by J. M. A. Lenihan, in "Mersenne and Gassendi," *Acustica* 1, no. 2 (1951): 96–99.

68   The other six members of the Accademia del Cimento were: Antonio Oliva (? –1668), the brothers Candido del Buono (1618–76) and Paolo del Buono (1625–62), the academy's secretary Lorenzo Magalotti (1673–1712), Alessandro Marsili (1601–69/71?), and Francesco Redi (1626–97/8?).

69 *Saggi di naturali esperienze fatte nell' Accademia del cimento*,
    269 pp., fol. (Florence: Giuseppe Cocchini, first issue of 1st ed.,
    1666OKU; the usual first edition is dated 1667)DPW; English
    translation by Richard Waller, *Essayes of Natural Experiments
    made in the Academie del Cimento* (London: Benjamin Alsop,
    1684): (a) *Saggi*, pp. 16-22, fig. 7; *Essayes*, pp. 10-12; (b) *Saggi*,
    pp. 244-45; *Essayes*, pp. 141-42; (c) *Saggi*, pp. 96-100; *Essayes*,
    pp. 50-52.

70 Vincenzio Antinori, *Notizie Istoriche relative All' Accademia Del
    Cimento*, pp. 108-267 in *Scritti editi e inediti*, vol. unico (Flor-
    ence: G. Barbera, 1868): (a) These experiments are described in
    two letters from Viviani, pp. 165-73, for the translations of which
    I am indebted to Jean P. Brockhurst and Rocco S. Narcisi; (b)
    see, for example, pp. 234-35.

71 Giovanni Targione-Tozzeti, *Atti e memorie inedite dell'Accademia
    del Cimento*, 3 vols. in 4 (Florence: G. Tofani stampatore, L.
    Carlierli, librajo, 1780).NN

72 A letter from John Wallis, quoted in Charles R. Weld, *A History
    of the Royal Society*, 2 vols. (London, J. W. Parker, 1848), 1:
    30 ff. Several good histories of the Royal Society are available,
    the earliest of which is Thomas Spratt, *History of the Royal
    Society of London for the Improving of Natural Knowledge*
    (London: J. Martyn and J. Allestry, 1st ed., 1667, 2d ed., 1702).

73 From the first Journal Book of the Royal Society (see preceding
    note).

74 For a general guide to the source material on the early academies,
    see Martha Ornstein, *The Role of Scientific Societies in the Seven-
    teenth Century* (New York: Columbia University Thesis, 1913;
    reprinted, Chicago: Chicago University Press, 1928, 1938).

75 Joanne-Baptista Duhamel, *Regiae Scientiarum Academiae His-
    toria*, Bk. 2, sec. 3, chap. 2, §x, p. 158 (Paris: Stephanum
    Michallet, 1698). This is said also to be on p. 169 of the 2d edi-
    tion (Leipzig, 1700), but I have not been able to examine this
    edition—and also on p. 161 of the *Secunda editio priori longa
    auctior* (Paris: Joannem-Baptistam Delespine, 1701).

76 "Sur la vitesse du son," *Histoire de l'Académie Royale des
    Sciences (Paris)* (1738), pp. 1-5; also Cassini de Thury, "Sur la
    propagation du son," *Mémoires de l'Academie Royale des Sci-
    ences* (Paris) (1738), pp. 128-46.

77 G. L. Bianconi, the second of *Due lettere di fisica al Sign. Maffei*
    (Venice: Simone Occhi, 1746), pp. 75-110.

78 C. M. de la Condamine, "Relation Abregee d'un Voyage fait
    dans l'intérieur de l'amérique méridional," *Mémoires de l'Aca-*

démie Royale des Sciences (Paris) (1745), pp. 391–492 [488].

79 William Derham, "*Experimenta & Observationes* de Soni Motu aliisque ad id attinentibus," *Philosophical Transactions of the Royal Society (London)* 26 (1708): 2–35. Summarized in English in the Philosophical Transactions *Abridgements*.

80 Sagredo, in a letter to Galileo dated 11 April 1615, in vol. 12, pp. 167–69 [168] of Favaro's edition of *Le Opere di Galilei Galileo* (see n. 48, above). I am indebted to Professor Paul G. Ruggiers of the University of Oklahoma for the translation of this letter. C. DeWaard (see n. 82, below) refers to the air being "aspirated" from the closed vessel in this experiment, but this interpretation is not supported by Sagredo's letter.

81 *Isaac Beeckman, Journal, 1604–1634*, fol. 356 recto, at p. 146 of vol. 3 (edited with an introduction and notes by C. DeWaard, 4 vols. (The Hague: Martinus Nijhoff, 1939).

82 Torricelli's claim to the barometer rests on his letter to Michelangelo Ricci dated at Florence, 11 June 1644: in Torricelli, *Collected Works*, Gino Loria and Giuseppe Vassura, eds. (Faenza: G. Montanari, 1919), 3:186–88, and translated on pp. 70–73 of Magie, *Source Book in Physics* (New York: McGraw-Hill, 1935). Cornelis DeWaard has shown conclusively, however, in *L'expérience barométrique* (Thouars [deux-sevres]: Imprimerie Nouvelle, 1936), that the *principle* of the barometer was well known before the date of Torricelli's letter to Ricci. Beeckman (see preceding note) had given the correct explanation of the raising of water by a pump as early as 1615, and a water-barometer form of the classic Torricellian experiment was exhibited by Berti at least as early as 1640.

83 DeWaard has also rescued this "talented mathematician" from the concealment of Kircher's latinizing. He identifies him, in *L'expérience barométrique*, pp. 102, 110, as one of a circle that included Galileo, Masotti, Torricelli, Magiotti, et al.

84 Athanasius Kircher, *Musurgia Universalis sive Ars Magna consoni et dissoni in X libros digesta*, vol. 1, Bk. 1, chap. 6, *Digressio*, pp. 11–13 (fol., vol. 1, Bks. 1–7, 690 pp., Rome: Ex Typographia Haeredum Francisci Corbelletti, 1650; vol. 2, Bks. 8–10, 462 pp. + chapter and subject indices, Rome: Typis Ludovici Grignani, 1650).DPW

85 Otto von Guericke, *Experimenta Nova (ut vocantur) Magdeburgica de vacuo spatio* (fol., 244 pp., Amstelodami: Apud Joannem Janssonium a Waesberge, 1672).ELC The manuscript itself was completed in March 1663. A German translation by F. Dannemann was published as no. 59 in Ostwald's *Klassiker der Exakten*

Wissenschaften. The first account of von Guericke's air pump was communicated privately to the Jesuit professor of Physics at Wurzburg, Kaspar Schott (1608-66), who published it in his *Mechanica Hydraulico-Pneumatica* (Herbipoli: Henricus Pigrin, 1657):DPW (a) *Experimenta*, Bk. 3, chap. 15, pp. 91-92; (b) ibid., Bk. 4, chap. 10, pp. 138-40.

86   Robert Boyle, *New Experiments Physico-Mechanicall, Touching the Spring of the Air, and its Effects*, Experiment 27, pp. 205-10 (1st ed.), pp. 105-10 (2d ed.), or pp. 103-08 (3d ed.) (Oxford: Printed by H. Hall, Printer to the University, for Tho. Robinson. 1st ed., 1660; 2d ed., 1662. Also London: Printed by Miles Flesher for Richard Davis, Bookseller in Oxford, 3d ed., 1682.)

87   Francis Hauksbee: (a) "An Experiment made at a Meeting of the Royal Society, touching the Diminution of Sound in Air rarefy'd," *Philosophical Transactions of the Royal Society* (London) 24 (1705): 1904; (b) "An Account of an experiment made at a Meeting of the Royal Society at Gresham College, upon the Propagation of Sound in Condensed Air. Together with a Repetition of the same in the open Field," ibid., pp. 1902-04. (c) "An Account of an Experiment shewing that actual Sound is not to be transmitted through a Vacuum," ibid., 26 (1709): 367-68.

88   R. B. Lindsay, "Transmission of Sound through Air at Low Pressure," *American Journal of Physics* 16 (1948): 371-77.

89   Kircher, *Musurgia Universalis* (see n. 84, above): (a) Vol. 2, Bk. 9, pp. 308ff.: Layout of tone wheels, pp. 312-19; Bells, pp. 336-39; Strings, pp. 339-42; Automatic hydraulic organ, pp. 311, 330-35, 342-44. See also the discussion (and translation of pp. 334-35) in H. G. Farmer, *The Organ of the Ancients*, pp. 21-22, 32-33, 165-67 (London: Wm. Reeves, 1931); (b) Vol. 2, Bk. 8, pp. 185ff.; (c) Vol. 2, Bk. 9, pp. 352-55; (d) Vol. 1, Bk. 6, prob. 3, pp. 450-51; (e) Vol. 2, Bk. 9, pp. 244-46; (f) ibid., pp. 243ff., 264, 297-302; (g) ibid., chap. 4, pp. 283-89; (h) ibid., chap. 3, pp. 293, 295-96, 303, 307; (i) ibid., pp. 277-78, 303-05; (j) ibid., p. 272.

90   Athanasius Kircher, *Phonurgia Nova sive Conjugium Mechanicophysicum Artis Naturae Paranympha Phonosophia Concinnatum*, . . . (fol., 229 pp., Campidonae: Rudolphum Dreherr, 1673):DPW (a) Bk. 1, sec. 1, 2, pp. 18ff., sec. 4, chap. 2, pp. 78-81, chap. 5, pp. 92-101, esp. 99; (b) Bk. 1, sec. 4, chap. 1, pp. 73-78; (c) Bk. 1, sec. 6, chap. 1, pp. 111-13; see also testimonial letters annexed to the "Preface to the Reader"; (d) Bk. 1, sec. 4, chap. 5, pp. 89-91; sec. 7, chap. 10, pp. 142-43, chap. 11, pp. 158-63; (e) Bk. 1, sec. 6, chap. 1, pp. 113-14; sec. 7, chap. 1,

p. 124; (f) Bk. 1, sec. 3, chap. 3, pp. 71–73; sec. 7, chap. 5, pp.
130–34; (g) Bk. 1, sec. 6, chap. 3, pp. 117–18.

91   Sir Samuel Morland, *Tuba Stentoro-Phonica, An Instrument of
Excellent Use, As well at Sea, as at Land; Invented and variously
Experimented in the Year 1670.* (fol. brochure, 14 pp., London:
Printed by W. Godbid, and are to be Sold by M. Pitt, 1672): (a)
Pp. 13–14; (b) Pp. 9, 11.

92   *The Ear of Dionysius* is discussed by W. C. Sabine in his article on
Whispering Galleries, published posthumously in *Collected Papers
on Acoustics,* ed. T. Lyman, pp. 255–76 [274–76] (Cambridge:
Harvard University Press, 1922). See also Kircher: in *Musurgia
Universalis* (n. 89, above), vol. 2, Bk. 9, Praelusio 3, pp. 291–93;
and the same text plus some misprints and a new drawing in
Phonurgia Nova (n. 90, above), Bk. 1, sec. 4, chap. 3, pp. 82–85.

93   Vol. 1, p. 284/268 of Leonardo's *Notebooks* (see n. 44, above).

94   Giambattista della Porta, *Magiae Naturalis, Libri XX,* in Bk. 22,
chap. 5, pp. 296–97 (editio princeps, sm. fol., 303 pp., Neapoli:
Horatium Saluianum, DDLXXXVIII [*recte* 1589]).^{DPW} There
were many later editions. See, for example, pp. 607–19 (12°,
Hanoviae: typis Wechelianis, 1619), or pp. 654–56 (Amstelodami:
Apud Elizeum Weyerstraten, 1664). The quoted passage is taken
from the English edition, *Natural Magick: In Twenty Books,* pp.
400–01 (sm. fol., 409 pp., London: Thomas Young and Samuel
Speed, 1658).^{DPW} This topic is not included in the first (1558)
edition, which contains only the first four books and was written
by Porta when he was only seventeen years old.

95   Francis Bacon, *Sylva Sylvarum* [1627] Century 3, §285, pp.
434–35 (see n. 46, above): see also *De Augmentis (Advancement
of Learning)* [1605, 1623] Bk. 5, chap. 2, p. 417, in vol. 4 (1858)
of the same Spedding, Ellis, and Heath edition of *The Works of
Francis Bacon.*

96   Narcissus Marsh, "An introductory Essay to the doctrine of
Sounds, containing some proposals for the improvement of
Acoustics," *Philosophical Transactions of the Royal Society
(London)* 14 (1683): 472–88. The author's name is misleadingly
given in the by-line and the index merely as *Bishop Narcissus.*

97   "Report of a letter received from N. Cassegrain of Chartres,"
Journal des Scavans, supplement (2 May 1672), pp. 76–80.

98   "Extract of a letter from Mr. John Conyers, of his improvement
of Sir Samuel Morland's Speaking Trumpet," *Philosophical Transactions of the Royal Society (London)* 12 (1678): 1027–29.

99   Richard Helsham, *Lectures in Natural Philosophy,* 5th ed. (published posthumously by Bryan Robinson), lect. 5, p. 75. (London:

1st ed., 1739; 2d ed., 1743; 4th ed., 1767; 5th ed., London: J. Nourse, 1777; a 7th ed., "with considerable additions," Philadelphia: P. Byrne, 1802).

100 Matthew Young, *An Enquiry into the Principal Phaenomena of Sound and Musical Strings* (8°, 203 pp., London: G. Robinson, 1784), LC pp. 49-56.

101 Sir Isaac Newton, *Philosophia naturalis principia mathematica* (Londini: iussu Societatis Regiae ac Typis J. Streator, 1687); ed. 2 auctior et emendatior [edited by Roger Cotes], (Cantabrigiae: 1713); editio ultima, (Amstelodami: sumptibus Societatis, 1714); 3d ed. [prepared under the direction of Henry Pemberton], (Londini: Gu. & J. Innys, 1726); translated into English by Andrew Motte (London: Printed for Benjamin Motte, 1729); a definitive revision of the Motte translation by Florian Cajori (Berkeley: University of California Press, 1934; reprinted 1946, 1947). All quotations used are taken from the Cajori edition. (a) Young's quotation from Newton, presumably translated freely from one of the Latin editions, is replaced here by the Cajori text, Bk. 2, sec. 8, prop. 50, p. 384.

102 Rene Descartes, *La Dioptrique*, Discourse 2; contained in *Discours de la methode pour bien conduire sa raison, et chercher la verité dans les sciences* (Leyden: I. Maire, 1637); also in *Oeuvres*, ed. C. Adam et P. Tannery, 12 vols. (Paris: L. Cerf, 1897-1913), vol. 6.

103 Pierre de Fermat, *Epistle cxii*, to C. de la Chambre, 1 Jan. 1662, in *Correspondance*, vol. 2 (1894), pp. 457-63 (458) of *Oeuvres de P. de Fermat*, 4 vols. (Paris: Par les soins de MM. Paul Tannery et Charles Henry, pub. Gauthier-Villars et Fils, 1891-96).

104 For guidance to relevant original sources, see, for example. E. A. Burtt, *Metaphysical Foundations of Modern Physical Science*, 2d ed., pp. 26ff., 46, 70, 214 (London: Kegan Paul, Trench, Trubner & Co., 1925; 2d ed., revised, New York: Humanities Press, 1951).

105 F. M. Grimaldi, *Physico-mathesis de lumine coloribus et iride* (pub. posth., ed. Hieronymus Bernia, Bononiae: Ex typographia haeredis V. Benatii, 1665); extracts translated in Magie's *Source Book* (see n. 82, above), pp. 294-98.

106 Sir Isaac Newton, *Opticks* (London: Printed for S. Smith and B. Walford, 1st ed., 1704; Latin ed. prepared by Samuel Clarke, with Queries 17-23 added, idem., 1706; 2d English ed., with all 31 Queries, London: W. & J. Innys, 1718; 3d ed., idem., 1721; 4th ed., "corrected by the author's own hand, and left before his death with his bookseller," London: W. Innys, 1730; paperbound reprint of 4th ed., with introduction by Sir Edmund Whittaker

and preface by I. Bernard Cohen, New York: Dover Publications, 1952). (a) Bk. 2, pt. 3, props. 12ff.; (b) For the action of Bodies on Light, see Bk. 3, Queries 1-5; (c) Bk. 3, Query 28. Quotation taken from the Dover edition.

107 Thomas Young, "On the Theory of Light and Colours," *Philosophical Transactions of the Royal Society (London)* 92 (1802): 12-48; "An Account of some Cases of the Production of Colours, not higherto described," ibid., pp. 387-97; "Experiments and Calculations relative to physical Optics," ibid. 94 (1804): 1-16.

108 Thomas Young, "Outline of Experiments and Inquiries respecting Sound and Light [in a Letter to Edward Whitaker Gray, Sec. R. S.]," *Philosophical Transactions of the Royal Society (London)* 90 (1800): 106-50; "Of the Analogy between Light and Sound," pp. 125-30; "The Fusion of Sounds [Beats]," pp. 130-33.

109 A. J. Fresnel, "Sur la diffraction de la lumière" (read to the Academy 29 July 1818) *Mémoires de l'Académie Royale des Sciences de l'Institut de France* 5 (1821-22): 339-475.

110 Christiaan Huygens, *Traité de la lumière* (presented to the Paris Academy, 1678) (Leide: P. van der Aa, 1690); also elegantly "Rendered into English" by Silvanus P. Thompson (London: Macmillan and Co., 1912).

111 Huygens, *Traité de la lumière;* this passage is excised from the translation in Magie's *Source Book* (n. 80, above), pp. 283-85.

112 Christiaan Huygens: (a) *Horologium Oscillatorium* (Parisiis: F. Muguet, 1673)DPW; for a critical summary, see Ernst Mach, *Science of Mechanics* (n. 31, above), chap. 1, pp. 192-226 of the MacCormack translation; (b) J. Wallis, "A Summary Account given by Dr. John Wallis of the General Laws of Motion," *Philosophical Transactions of the Royal Society (London)* 3 (Nov. 26, 1668): 864-66; Sir Christopher Wren, "Lex Naturae de Collisione Corporum," ibid. (Dec. 17, 1668), pp. 867-68; C. Huygens, "A Summary Account of the Laws of Motion," ibid. 4 (Jan. 4, 1669): 925-28, and also *De Motu Corporum ex Percussione in Opuscula Posthuma* (Leyden, 1703).

113 Robert Hooke: the first publication of "Hooke's law," in the form of an anagram, appears in *A Description of Helioscopes, and some other Instruments*, p. 31 (London: Printed by T. R. for John Martyn, 1676); the anagram is explained, and it is claimed that the original discovery had been made in 1660, in *Lectures De Potentia Restitutiva, or of Spring* (London: Printed for John Martyn, 1678); these are combined with other tracts in *Lectiones Cutlerianae, or A Collection of Lectures, Physical, Mechanical, Geographical, and Astronomical. Made before the Royal Society*

on several Occasions at Gresham College. To which are added divers Miscellaneous Discourses. (London: Printed for John Martyn, 1679).DPW

114   The Diary of Robert Hooke, 1672-1680, ed. Henry W. Robinson and Walter Adams (London: Taylor & Francis, 1935), entries for: (a) Saturday, January 15, 1675/76; (b) Tuesday, March 28th, 1676.

115   Richard Waller, "The life of Dr. Robert Hooke," p. xxiii, prefixed to The Posthumous Works of Robert Hooke (London: Sam. Smith and Benj. Walford, 1705).DPW

116   Felix Savart, "Notes sur la sensibilité de l'organe de l'ouie," Annales de Chimie et de Physique [2e] 44 (1830): 337-52.

117   Hooke, Posthumous Works (see n. 115, above), in "The Method of Improving Natural Philosophy," pp. 39-40.

118   William Harvey, Exercitatio anatomica de motu cordis et sanguinis in animalibus (Francofurti: Sumptibus Guilielmi Fitzeri, 1628); the quotation is from the modern English translation by Chauncey D. Leake, 3d ed. (Springfield, Ill. and Baltimore, Md.: Charles C. Thomas, Pub., Tercentennial facsimile ed., 1928; reprinted 1931, 1941), p. 49.

119   See G. Joseph, "Geschichte der Physiologie der Herztöne vor und nach Laënnec bis 1852," Janus [Central-Magazin für Geschichte und Literargeschichte der Medicin, . . .] 2, nos. 1-3 (new series, Gotha, 1852; reprinted, Leipzig: Alfred Lorentz Buchhandlung, 1931).LC

120   R. T. H. Laennec, De l'auscultation médiate, subtitled "A Treatise on the Diseases of the Chest, in which they are described according to their Anatomical Characters, and their Diagnosis established on a new principle by means of Acoustick Instruments." Translated from the French of R. T. H. Laennec, M.D., with a preface and notes by John Forbes, M.D. (1st American edition, Philadelphia: James Webster, 1832).LC The quotation is from Part Second (Diagnosis, Introduction), p. 211. Drawings of the stethoscope and details of construction are given in the 26e Leçon (Lesson 26) of Laennec's De l'auscultation médiate, 2 vols. (Paris, J.-A. Brosson et J.-S. Chaude, 1819). I am indebted to Professor Osman K. Mawardi for uncovering the latter reference.

Appendix

1   E. F. F. Chladni, Die Akustik, preface (Leipzig: Breitkopf und Haertel, 1802).

2   A. Conan Doyle, Memoirs of Sherlock Holmes, chap. 1, "Silver Blaze." Many editions. For example, vol. 1, sec. 4, p. 23 of The

Complete Sherlock Holmes, Memorial ed. (New York: Doubleday, Doran and Company, 1930).

3   Sir Francis Bacon, *De Augmentis* (Advancement of Learning), Book 3, chap. 6, p. 371, in vol. 4 of *The Works of Francis Bacon*, ed. J. Spedding, R. L. Ellis, and D. D. Heath (London: Longman and Co., et al., 1858).

4   William Thomson (Lord Kelvin), in a lecture entitled "Electrical Units of Measurement," announced for 3 May 1883 and published in *The Practical Applications of Electricity*, p. 149 (London: Published by the Institution [of Civil Engineers], 1884).

5   See, for example, Carl B. Boyer, *Concepts of the Calculus* (New York: Columbia University Press, 1939); also, W. W. R. Ball, *A Short History of Mathematics*, 4th ed. (London: Macmillan and Co., 1908; 6th reprinting, 1924), chaps. 16 and 17.

6   "A Letter of Mr. Isaac Newton, Professor of the Mathematicks in the University of Cambridge; containing his New Theory about Light and Colors: sent by the Author to the Publisher from Cambridge, Feb. 6, $16\frac{71}{72}$; in order to be communicated to the R. Society." *Philosophical Transactions of the Royal Society (London)* 6, no. 80 (February 19, 1671/72): 3075–87.

7   Laplace, *Précis de l'histoire de l'astronomie* (Paris: Veuve Courcier, 1821), p. 110; the same text forms Book 5 of the 5th edition of *Exposition du Système du Monde* (Paris: Bachelier, 1824).

8   Quoted from [Jean Baptiste Joseph] Delambre's obituary *Notice sur la vie et les ouvrages de M. Malus et de M. le Compte Lagrange*, pp. xxvii–lxxx (xlvi), *Histoire de la classe des Sciences Mathématiques et Physique de l'Institut Royal de France*, 1812.

9   Letter from Newton to Robert Hooke, dated "Cambridge, February 5, 1675-6"; quoted in Sir David Brewster, *Memoirs of the life, writings, and discoveries of Sir I. Newton*, 2 vols. (Edinburgh: T. Constable and Co., 1855), chap. 6, p. 142. For other general biographical material on Newton, see also: Brewster's *The Life of Sir Isaac Newton* (London: J. Murray, 1831); Louis Trenchard More, *Isaac Newton, a biography* (New York and London: Charles Scribner's Sons, 1934); Edward Neville da Costa Andrade, *Isaac Newton* (London: Parrish, 1950); and I. B. Cohen, *Franklin and Newton: An Inquiry into Speculative Newtonian Experimental Science and Franklin's Work in Electricity as an Example Thereof* (Philadelphia: The American Philosophical Society, 1956).

10   D'Alembert, *Essai d'une nouvelle théorie de la résistance des fluides*, introduction (Paris: Chez David l'aîné, 1752), pp. xiv–xv.

11   Newton's calculation of sound speed appears in the *Scholium* sub-

    joined to Prop. 50. He did not identify the sources for the experimental speeds he quoted and the speeds do not coincide with the results of any of the sound-speed experiments known to have been carried out before 1687.

12 Mairan, "Discours Sur la Propagation du Son dans les différents Tons qui le modifient," *Mémoires de l'Académie Royale des Sciences (Paris)* (Paris, 1737), pp. 1-58 [1-85, Amst.]; noticed in the Editor's "Diverses Observations de Physique générale," *Histoire de l'Académie Royale des Sciences (Paris)* (Paris, 1720), pp. 11-12 [14-15, Amst.].

13 I. Newton, *Opticks* (see chap. 2, n. 106, above), Book 1, pt. 2, Prop. 3, Prob. 1, Exp. 7, pp. 126-28; also Book 2, pt. 2, Observation 14, p. 212.

14 Sir John Hawkins, *A General History of the Science and Practice of Music*, 5 vols., 5:67-69 (London: Printed for T. Payne and son, 1776; a new ed., with the author's posthumous notes, 2 vols., London, New York: J. A. Novello, 1853; reprinted in 3 vols., London: Novello, Ewer & Co. and New York: J. L. Peters, 1875).

15 A. A. Michelson, *Light Waves and Their Uses* (Chicago: University of Chicago Press, 1903; reprinted 1907), p. 2.

16 Sir J. J. Thompson, *Nature 119*, Supplement, no. 2995 (26 March 1927): 36-40.

17 Euler, *Dissertatio physica de sono*, 16 pp. (Basel, 1727), and in *Euleri Opera Omnia*, ser. 3, vol. 1, pp. 182-96: a critical synopsis in the editor's [C. A. Truesdell] introduction to *Euleri Opera Omnia*, ser. 2, vol. 13, pp. 18-23. See also Euler, "Tentamen explicationis phaenomenorum aeris," *Commentarii acad. sci. Petropolitanae* 2 [1727]: 347-68 (1729); synopsis in Truesdell's introduction to *Euleri Opera Omnia*, ser. 2, 12:xxi-xxii.

18 Lambert, "Sur la vitesse du son," *Mémoires de l'Académie Royale des Sciences (Berlin)* 24 [for 1768]: 70-79 (1770); "Sur la densité de l'air," pp. 48-56, *Nouvelles mémoires de l'Académie Royale des Sciences* (Berlin) 1772 (1774): 103-40.

19 Lagrange, "Nouvelles recherches sur la nature et la propagation du son," *Miscellanea Taurinensia* 2 [pt. 2, for 1760-61] (1762): 11-172.

20 The "ingenious gentleman Mr. Richard Townley" mentioned by Boyle seems not to have been identified further by most of the science historians who have noticed him, perhaps because there were at least six branches of the Towneley family, each of which had a Richard in almost every generation. The Richard Towneley, Esq. (sometimes Townley) whose dates are given here was a retiring and studious amateur scientist who lived principally at

Towneley in Lancashire; he was the oldest son of Charles Towneley, was married to Margaret Paston, died "of a mortification" in York, and was buried in Burnley. See Thomas Dunham Whitaker, *An History of the Original Parish of Whalley* . . . (Blackburn: Printed by Hemingway and Crook, 1800; 4th ed., 1876), Book 6, chap. 1, p. 465.

21   *A Defence of the Doctrine touching the Spring and Weight of the Air, proposed by Mr. R. Boyle, in his New Physico-Mechanical Experiments; against the Objections of Franciscus Linus. Wherewith the Objector's Funicular Hypothesis is also examined;* contained in the 2d edition (1662) of Boyle's *New Experiments Physico-Mechanical* . . . (see chap. 2, note 86, above). Also in the collection of *The Works of the Honourable Robert Boyle*, 6 vols. ("A new edition," London: printed for J. and F. Rivington . . . , 1772), 1:156-59.

22   Mariotte, *De la nature de l'air* (Paris: E. Michallet, 1679).

23   Amontons, "Moyen de substituer commodément l'action du feu à la force des hommes et des chevaux, pour pouvoir les Machines," *Mémoires de l'Académie Royale des Sciences* (Paris), 1699: 112-34 [114] (Amst. 1699: 2:159-79 [162]); see also, "Discours sur quelques propriétéz de l'air, et le moyen d'en connoître la température dans tous les climats de la terre," ibid., 1702, *Histoire*, pp. 1-8, *Mémoires*, pp. 155-74; parts of the latter are translated in Magie, *Source Book in Physics* (New York: McGraw-Hill Book Co., 1935), pp. 128-31.

24   Wm. Thomson (Lord Kelvin), "On an Absolute Thermometric Scale . . .", *Proceedings of the Cambridge Philosophical Society* 1 (1848): 66-71.

25   Amontons, "Le Thermomètre . . .", *Mémoires de l'Académie Royale des Sciences (Paris)* 1703: 50-56 [*Amst.* 64-72].

26   For a general discussion of the evolution of temperature scales, see H. C. Bolton, *The Evolution of the Thermometer, 1592-1743* (Easton, Pa.: The Chemical Publishing Co., 1900), and Carl B. Boyer, "History of the Measurement of Heat," *Scientific Monthly* 57 (1943): 442-52, 546-54.

27   Lambert, *Pyrometrie oder vom Maase des Feuers und der Wärme* (Berlin: ben Haude und Spener, 1779), p. 29.

28   Gay-Lussac, "Recherches sur la dilatation des gaz et des vapeurs," *Annales de Chimie* 43 (1803): 137-75.

29   John Dalton, "Experimental Essays . . . (Essay IV) on the Expansion of Gases by Heat," *Memoirs of the Literary and Philosophical Society of Manchester* 5, pt. 2 (1802): 595-602 [600].

30   The seven pioneer energy conservationists were Carnot, Mayer, Holtzmann, Joule, Seguin, Mohr, and Colding. For a brief ac-

count of the origins of this principle, written while the controversy was still fresh, see L. A. Colding, "On the History of the
Principle of the Conservation of Energy," *Philosophical Magazine*
27 (1864): 56-64.

31 This point is tellingly made by Thomas S. Kuhn in his incisive
paper "The Caloric Theory of Adiabatic Compression," *Isis* 49
(1958): 132-40. I am indebted to Professor Kuhn for early access
to this manuscript and for guidance to the primary literature on
adiabatic effects.

32 Sadi Carnot, "Réflections sur la puissance motrice de feu et sur
les machines propres a développer cette puissance" (Paris: Chez
Bachelier, libraire, 1824).

33 Poisson, "Sur la chaleur des gaz et des vapeurs," *Annales de
chimie et de physique* 23 (1823): 337-52.

34 William Cullen, "Of the Cold produced by Evaporating Fluids,
and of some other Means of producing Cold," *Essays and Obser-
vations, Physical and Literary* [of the Philosophical Society of
Edinburgh] 2: 145-56 [153] (Edinburgh: G. Hamilton and J.
Balfour, 1756), or pp. 159-71 [168] of the 2d ed. (Edinburgh:
Printed for John Balfour, 1770).

35 J. C. Arnold, *Erlangische Gelehrte Anmerkungen und Nach-
richter XLIX Stück* (1759), pp. 436-40.

36 J. H. Lambert, *Pyrometrie oder vom Maase des Feuers und der
Wärme* (Berlin: ben Haude und Spener, 1779): (a) § 493, pp.
266-67; (b) § 495, p. 268.

37 Horace Bénédict de Saussure, *Essais sur L'Hygrométrie* (Neu-
chatel: Samuel Fauche père et fils, 1783).

38 Marc Auguste Pictet, *Essais de Physique*, vol. 1: Sur le feu
(Geneva: Chez Barde, Manget & Compagnie, 1790).

39 Erasmus Darwin, "Frigorific Experiments on the Mechanical Expansion of Air," *Philosophical Transactions* 78 (1788): 43-52.

40 L. A. von Arnim, *Annalen der Physik* 2 (1799): 238-45.

41 John Dalton, "Experiments and Observations on the Heat and
Cold produced by the Mechanical Condensation on Rarefaction
of Air," *Memoirs of the Literary and Philosophical Society of
Manchester* 5 (1802): 515-26.

INDEX